Copyright © 2022

All rights reserved.

ISBN:
ISBN-13:

How to Write a Graduate Student Fellowship

by Rebecca M. C. Spencer, PhD
University of Massachusetts, Amherst
winningsleep@gmail.com

CONTENTS

1: Graduate Fellowship Proposals: Introduction ... 1
    Why write a graduate fellowship proposal? ... 1
    Who is qualified for the NIH and NSF graduate research fellowships? ... 2
    So what should I apply for? ... 3

2: Getting Started ... 7
    Identifying your idea ... 7
    Mind the gap ... 8
    Developing your goal into an overarching objective and specific aims. ... 9
    Checking your aims ... 12
    Define your hypotheses ... 13

3: Testing Your Scope ... 15
    Sketching your approach ... 15
    Checking your approach ... 16
    Sketching out your timeline ... 16
    Checking your timeline ... 17

4: Some Advice on Good Writing ... 19
    Making time to write ... 19
    Tips and tricks for good writing ... 20

Section 2: Applying for the NIH F31 (NRSA)
5: Specific Aims ... 28

- Interest grabbing statement, framing in health relevance 28
- The long-term goal 30
- The impact paragraph ..... ??
- Pulling it together ..... ??

6: Selection of Sponsor and Institution 32
- The goal of the Selection of Sponsor and Institution section 32
- Writing the Selection of Sponsor and Institution section 32

7: Research Strategy 36
- Overview of the Research Strategy 36
- Significance (1-1.5 pages) 36
- Innovation (.3-.5 pages) 39
- Approach (2.75-3.5 pages) 41
- Potential Problems and Alternative Strategies (.3 pages) 45
- Feasibility 46
- Timeline and Benchmarks for Success (.25 pages) 46

8: Training in Responsible Conduct in Research 47
- The goal of the Responsible Conduct in Research section 47
- Writing your Responsible Conduct in Research section 47

9: Biosketch 51
- What is a Biosketch? 51
- Header information 51
- Education and training 52
- Personal statement 53
- Positions and honors 55

- Contributions to science 56
- Additional information: Scholastic performance 57

10: Applicant's Background and Goals for the Fellowship 59
- The goal of the Background and Goals section 59
- Writing your Background and Goals section 60

11: Facilities and Other Resources 65
- The goal of the Facilities and Other Resources section 65
- Writing your Facilities and Other Resources section 65

12: Letters of Support, Recommendation, and Eligibility 69
- Types of letters 69
- Letters of reference 69
  - The goal of the Letters of Reference 69
  - Who should letter writers be? 71
  - Requesting letters of reference 73
  - Instructions for letter writers 74
- Letters of support 77
  - The goal of the letter of support 77
  - Who should provide a letter of support? 77
  - Drafting the letter of support 77
- Letter of eligibility .... ??
  - The goal of the letter of eligibility .... ??
  - Who should provide the letter of eligibility? .... ??
  - Drafting the letter of eligibility .... ??

13: Sponsor/Co-sponsor Requests 79

- Materials needed from the sponsor/co-sponsor ... 79
- Sponsor/Co-sponsor Biosketch ... 79
- Sponsor/Co-sponsor Statement ... 80

14: Institutional and Environmental Commitment ... 83
- The goal of the Institutional and Environmental Commitment section ... 83
- Drafting the Institutional and Environmental Commitment section ... 83

15: Respective Contributions ... 86
- The goal of the Respective Contributions section ... 86
- Writing the Respective Contributions section ... 87

16: Proposal Summary/Abstract ... 89
- About the Proposal Summary ... 89
- Writing the Proposal Summary ... 89

17: Human Subjects ... 92
- Protection of Human Subjects section ... 92
- Writing the Protection of Human Subjects section ... 92

18: Vertebrate Animals ... 96
- Vertebrate Animals section ... 96
- Writing the Vertebrate Animals section ... 96

19: Resource Sharing Plan ... 98
- About the Resource Sharing Plan section ... 98
- Writing the Resource Sharing Pllan ... 98

20: Bibliography/References Cited ... 100
- Formatting the bibliography ... 100

21: Editing your Proposal ... 102
    Seeing the Proposal through Reviewer's eyes ... 102

CHAPTER ONE

Graduate Fellowship Proposals: Introduction

Chances are, if you have picked up this book, you do not need to be convinced that writing a graduate fellowship proposal is something you want to do. But maybe you are on the fence about applying or are part of a course where someone (your PI, another professor, or your graduate program) is <u>making</u> you apply. It is helpful to find internal motivation to get you through the rather laborious process. So here are some reasons to consider.

Reasons to apply for a graduate fellowship

To get a paycheck. If you are paying your own way through your graduate training, tuition is one thing but the lack of a paycheck is quite another. A graduate fellowship typically covers both your tuition as well as a stipend that goes back to you (a paycheck!). The pay may not be great (see **Box 1**), but perhaps better than pinching pennies.

To dedicate more time to your research. Perhaps you get a check but that comes from the substantial amount of time you spend either in an on- or off-campus job or by teaching (via a teaching assistantship). While some of these experiences may benefit your career, they may take time away from <u>your</u> research objectives which may compromise the quality and speed of that work. With your tuition covered and a stipend provided, typically graduate fellows do not teach or work in other jobs.

To improve your CV. The old saying goes, "getting funding gets you funding." Okay, it might not be that old of a saying and it may be said in small closed circles, but it is a concept that is supported in academia: having successfully received a grant becomes a marker of success waived on your next grant application which improves its likelihood of success and so forth. Also, employers (including faculty looking to hire a post-doc, faculty search committees looking to hire a junior faculty member) are consistently looking for a history of self-funding. So, overall, grants will improve your CV, future job prospects, and career.

To get experience grant writing. The process of grant writing is an art. But many aspects of writing a graduate fellowship application will translate to writing a post-doctoral fellowship application and faculty grants someday.

To develop your research ideas. The exercise of grant writing forces you to process and plan your research in a way you might not otherwise do. Most students report being better prepared for their dissertation process (the proposal through defense stages) by having prepared a fellowship application on the work.

So, whether funded or not, the process pays off.

Box 1. How much is a fellowship worth?
The value of a graduate fellowship may change year-to-year so see the relevant funding announcement to get the most up-to-date numbers. Currently (spring 2022), the value is:

	Number of Years	Stipend (yearly)	Cost-of-Education/ Tuition (yearly)	Other (yearly)
NSF Graduate Research Fellowship	3 years	$34,000	$12,000	None
NIH National Research Service Award	1-5 years	$26,353	Up to $21,000	Research Allowance: $4,400
Ford Foundation	Min 3 years	$27,000	none	none

Before getting further, it is probably a good idea to check that you are eligible for one of the awards focused on in this book - the NSF Graduate Research Fellowship and the NIH National Research Service Award. In addition to assuring *you* fit the eligibility criteria, be sure that your *research* fits the mission of one and/or the other funding agency.

* * *

	NSF Graduate Research Fellowship	NIH National Research Service Award (F31)
Eligible candidates	• Must be a U.S. Citizen or permanent resident • Not have previously been awarded an NSF GRFP • Not have a graduate degree (i.e., cannot have a Masters degree) • Must apply in the year prior to your third year of the graduate program (prior to starting grad school or your 1st or 2nd year of grad school). Can apply only once as a graduate student.	• Must be a U.S. Citizen or permanent resident • Enrolled in a PhD program in the biomedical, behavioral, or clinical sciences • In the dissertation research stage • Evidence of "high academic performance in the sciences, and commitment to a career as an independent scientist"
Eligible research	NSF funds fundamental research and education in all the *non-medical* fields of science and engineering. NSF areas include: Geosciences Life sciences Computer and information sciences Engineering Materials research Psychology Social sciences STEM education and learning Chemistry Math Physics and astronomy	NIH funds research in scientific health-related fields relevant to the individual institutes/centers' missions. NIH areas include: Cancer research (NCI) Vision research (NEI) Heart/lung research (NHLBI) Genetics research (NHGRI) Aging research (NIA) Alcohol abuse research (NIAAA) Allergy/infectious disease (NIAID) Musculoskeletal and skin diseases (NIAMS) Biomedical/bioengineering (NIBIB) Human development (NICHD) Speech/hearing research (NIDCD) Dental/craniofacial research (NIDCR) Digestive and kidney diseases (NIDDK) Drug abuse research (NIDA) Environmental health. (NIEHS) General medicine (NIGMS) Mental health research (NIMH) Health disparities (NIMHD) Neurological diseases/strike (NINDS) Nursing research (NINR)

If you are not eligible for these, keep in mind that there are other graduate research funding opportunities out there to consider. For instance, check out the National Defense Science and Engineering Grants, the Ford Foundation Fellowships, American Heart Association, or the Department of Energy Fellowships. Your institution may also have grants to apply for to support graduate research.

Most graduate students in STEM fields will qualify for the NIH or NSF award as long as they are a U.S. citizen and apply early enough. And most will qualify for only one of these awards, with research qualifying either as a non-medical field (and therefore NSF worthy) or a medical/health-related field (and thus NIH worthy). However, some research can be spun for either, in which case, **applying to both is strongly encouraged**. In this case, it is often best to start with the NSF award because, simply, the application is less intensive and time-restricted (must be applied for by your 2nd year or earlier). After that, pursue the NIH award for the subsequent available cycle. Likewise, a student eligible for the Ford Foundation fellowship is encouraged to apply for that in addition to a NSF or NIH application.

While you can generally only accept one award (and certainly one at a time), there are benefits to applying for more than one.
   1) Obviously this will increase your chances of something getting funded.
   2) You will learn more about the grantsmanship process as you make adjustments to target the specific funding agency.
   3) You will continue to learn about your research through this process.
   4) If you are so amazing that you are awarded more than one, you can decline one of the awards but still list this on your CV/Biosketch (such as "NIH F31 5555555 *Title of the work* - awarded but declined")

A common question is when should you apply for a graduate fellowship. On the one hand, earlier funding may allow you the support for more years of your graduate career. It is also important to apply early as you will more than likely need to resubmit and this process takes time. On the other hand, if you apply too early, your ideas may not be developed enough and you may not have sufficient publications/experiences and you are unlikely to be successful.

When is the right time depends in part on the fellowship you are applying to. NSF is rather straightforward while NIH is much more nuanced.

When to apply for an NSF GRFP
NSF grad fellowships are only available for applicants prior to grad school (applying in your final year of undergrad) or in your first or second year of graduate school. Reviewers are instructed to scale their expectations to career level. This of course is a subjective assessment as it is and scaling something that is really not scalable (because those expectations - things like publications, conference presentations - will also vary by research area) is impossible, so take this expectation very lightly.

* * *

Applying before grad school is rare because it is hard. It's a busy time and often students are focused on getting into grad school and where to go to grad school, that applying for an NSF fellowship is not on the radar. It is also tricky to have research to propose before you've matched with a graduate mentor and program. But, surprisingly to most, a match is not needed to apply! You can write a propose about potential research without regard to who you would do it with or where. Nonetheless, most senior undergrads could not come up with a research idea without the support of a mentor. If you have a great idea as an undergraduate, apply! The good news is you can apply again as a graduate student.

If you did not apply as an undergraduate, you can only apply once - either in your first or second year of graduate school. So which is better? For this, you should assess how prepared you are. Use the following flow chart to help you decide.

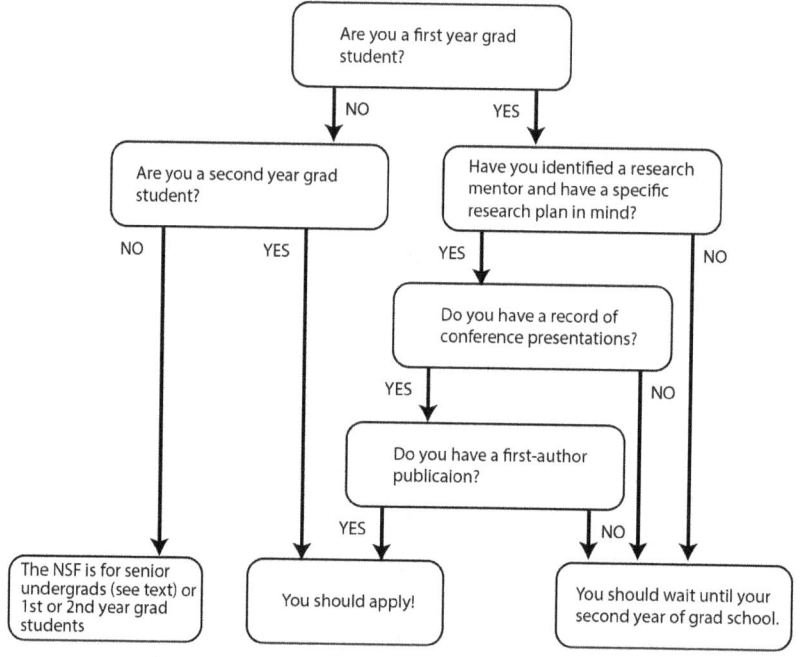

When to apply for an NIH F31

NIH has no restrictions on which year you are in your graduate program when you apply. They only require that you are "in the dissertation research stage of training" which, for many programs, is a bit vague.

NIH reviewers do consistently want to see that you have a publication or even more than one at the time of your application. Critiques suggest that they also like to see some experience in the area of your research proposal, meaning, it may help if the publication is in the same research area as your proposal.

These things suggest applying later is better. However, for NIH, it is important to keep in mind, that few get funded on the first submission but re-submissions are allowed. With the typically-detailed feedback from reviewers, it is worth reapplying. So it is worth considering a timeline that also let's you reapply. Here is an example timeline. If you submit a grant in Dec, you will have a score by March but a funding decision not until May. So, assuming you need to reapply, the next option (assume you missed the April 8 deadline) is Aug 8. That submission will not be scored until around November with funding decisions in January. So this process can take about 1 year and should be factored in your decision as to when to apply. My general recommendation is to apply for the first time around your second year of your graduate training.

CHAPTER TWO
Getting Started

You entered graduate school with a goal of getting a degree. You knew at least what you wanted that degree to be in – Anthropology, Neuroscience, Chemistry, etc. But you probably knew even more than that. You knew, for example, you wanted to be in <u>Cultural</u> Anthropology, <u>Cognitive</u> Neuroscience, or <u>Medicinal</u> Chemistry. What's more, by now, you have likely identified a specific research group and mentor you want to work with. See that, you're off to a great start!

How do you identify <u>your</u> research idea having narrowed down your research group? Here's where you have to do some legwork. Consider the following:

- <u>Find what interests you.</u> Something brought you to this research group in the first place. What was interesting to you about the work of this research group? Stay centered in that because, let's face it, you want to work on something you are innately interested in.
- <u>Know your tools.</u> Your research group provides your toolbox. The group uses some array of techniques or approaches to questions that provide the tools in your toolbox. Outside your research group, what other questions have been answered with these tools? Now, you may end up with a question that you need unique tools to answer so don't feel limited by the tools at your disposal, but certainly know what is in your toolbox before reaching into someone else's toolbox.
- <u>Most importantly, know the foundation that you stand on.</u> To develop your own idea, you must know what has already been done. Think of it this way: If you build a house, it must be on a solid foundation. Likewise, in research, the foundation is important. The

foundation is, of course, all of the research in that area that has come before you. It will take some time wandering around on that hypothetical foundation before you see "the cracks" (see **Box 2**). These 'cracks' are the gaps in the literature. What assumptions are untested that your research might contribute to? Or perhaps this foundation would be useful for other buildings to be built on. What unique applications of the concept do you see that you might contribute to?

Box 2. The Foundation: How to learn the literature without being overwhelmed.

If you are new to a research area, it is easy to be overwhelmed by the number of papers you have to read to feel like you can speak up in a presentation let alone pose your own research question. In most fields, you cannot possibly read all of the literature before you get a project underway. However, your graduate career is the point when you will have most opportunity to read papers, so you should make every effort to constantly learn what was out there when you stepped in the door but also what is being published in your field every single day. Here are some pointers:

- **Narrow down your field.** To hit the ground, you do not need to know the broad field that you are a part of. Start by knowing the sub-field. For some areas, even this can be overwhelming, so perhaps narrow it down further. For example, my research area may be Psychology, my field may be Cognitive Psychology, and my sub-field may be "Memory". But "Memory" is still a huge sub-field with research dating back over 150 years! So focus on something more narrow – The neural representation of memory? Age-related changes in memory? Etc.
- **Once you've defined a narrower field, where do you start?** One place to start is by reading (and knowing really, really, really well!) the work from your research group. As you read these papers, where do they take you? Note which references seem critical to the research and seek these out. These papers will then have other critical references, and so forth.
- **Don't overlook the other point of view.** Sometimes a research sub-field can be split into two (or more) camps. These camps tend to focus on their own beliefs and fail to cite the work in the other camps. Be sure to seek out research in the other camp. Understand why the split exists. Ask questions (particularly inside your research group) if you do not understand the rift or your research group's position.
- **Keep up on current literature.** In addition to playing catch-up on the literature, it's also important to keep up with the constantly evolving work in your field. The best way to do this is to sign up for alerts for certain search terms using Google Scholar or other format. Also, follow your journal, conference, and labs in your field on social media.

Putting those things together, hopefully you can find **something of interest**, using **tools available**, to **fill an important gap** in your research subfield! But a word of caution on 'filling a gap'.

It is often easy to find a gap in the literature. The thinking is often: *"They found this by testing X. I will see if they find the same thing by testing Y."* While that is certainly an identified gap, *not all gaps need to be filled*. Or, at least, filling a gap is not always fundable research. It is important that the gap you seek to fill be a necessary gap to be filled. Back to the analogy to a foundation, if there is a crack in the foundation, but it is far outside of the area where you place the house or is microscopic and not going to affect the stability of the house, then you aren't going to get a funder to give you money to fill it (reviewers love the word "incremental"!). So, keep in mind that you will need to justify your research and consider alternative hypotheses (more to come on that).

Also, the gap must not be <u>too narrow</u>. Small holes may not need to be filled (lack significance), but also, they will not be sufficient to define the start of your career. You need a project that will be a sufficient scope of a dissertation and springboard for your career.

The gap must also not be <u>too vast</u> but if it is, you may at least still have an idea that you can narrow down to an appropriate scope that is 'solvable' in the scope of a dissertation.

> At this point, hopefully you have defined the gap. Write out a **statement of the gap** you hope to fill. For example: *"We know odors can aid learning in young adults but we do not know if they aid memory in older adults."*

> Also, check that this gap is an important and necessary gap to be filled by writing out a **statement of need**. For example, *"Understanding whether odors aid learning in older adults would provide the basis for a practical memory intervention for older adults."*

You have an idea! That's a great start. Now you need to make that idea into an **overarching objective** with a few **specific aims**.

[Note: This is where talking to a broad audience of PhD students gets dicey because the scope of dissertation research can vary tremendously from one area to the next. It is always important to seek out examples of successful fellowships from your field but also dissertations from your research group and others in your division. These dissertations will give you a sense of the scope of work expected for a degree and that is typically what is also expected by fellowship application reviewers from those fields.]

Your **overarching objective** may very well be to fill the gap you identified. This statement should clarify the thread that ties the specific aims together (more to come on that).

<u>A common mistake here</u> is that students make a statement of

overarching objective that is really the statement of two more more specific aims in one sentence. For instance, they might state "The overarching objective of the proposed research is to understand whether an odor cue enhances memory in older adults <u>and</u> whether memory changes in older adults differ from young adults." You can guess what the two aims are going to be because this is not a single overarching objective, it is the statement of the two specific objectives in one sentence.

> Now write out a **statement of your overarching objective**. If you defined your gap, you can start by rewording that. For example, *"The overarching objective of the proposed research is to understand whether an odor cue can enhance memory in older adults."*

Your **specific aims** (often called 'specific objectives' in NSF lingo) are critical to your plan. More than anything else, this is what is seen as your commitment to what you will get done.

Typically, multiple aims are required for a number of reasons:
- To sufficiently address the overarching hypothesis, it may need a multi-pronged approach or a series of studies to make headway towards this goal
- A thesis project often requires a scope large enough that cannot be achieved in a single project/study, so to achieve the appropriate scope, a set of experiments, with each often representing an aim, is often (but not always) needed.

How do you carve out specific aims once you've defined the gap and overarching objective? Consider your overarching objective and what study would be conducted to achieve it. This, in itself, will be one specific aim. **Take the time to write it as such.** Meaning, an aim is most often stated as "To examine…" or "To characterize…" (see below for an example). Re-read it. Conduct a series of checks:
- Is this aim <u>achievable</u>? You may find that wording such as "To understand…" may end up with a less achievable objective than "To examine" (or "to determine". An examination could be completed in the next few years whereas an understanding suggests possibly more and will a complete understanding of the issue be achieved in that amount of time? You may even have it phrased as "To examine…" but, when you consider the scope of the statement, it may still be unattainable. Consider rewording this to an achievable objective. In some cases, as you do this, you may

identify the second aim that will be needed to have a fully developed proposal.
- Is the outcome <u>too obvious</u>? As we will consider later, the majority of research is hypothesis-driven and will have a hypothesis (and should have a reasonable alternative hypothesis). If you consider research in the field and study X, study Y, and study Z all find a certain outcome and you are expecting the same outcome in your study, then ask yourself whether this study is necessary, will it have impact, and can you convince reviewers that it is nonetheless interesting and should be funded?

Still haven't identified two (or more) aims? Here are a few approaches to consider:
- **Consider whether your overarching aim could be broader.** Perhaps what you have written as an overarching aim should instead be a specific aim and you could achieve more. What would be a reasonable parallel question and what might be the overarching aim that this would accomplish? For instance, perhaps my overarching aim should not be *"to understand whether odor cues can enhance memory in older adults."* Perhaps that should become Aim 1 and my Aim 2 be *"to understand whether auditory cues can enhance memory in older adults"* thus making my revised overarching objective *"to understand whether contextual memory cues can enhance memory in older adults."*
- **Consider whether your question can be approached from multiple angles.** Here, say to yourself – if my overarching objective is true, what are 2 (or 3 or 4) other results that I would expect. Or, perhaps you could say that your overarching hypothesis should be true in 2 (or 3 or 4) domains, but that it is necessary to test each as differences in domains have been shown elsewhere. An example of this is below. In the area of memory, there are differences between declarative and procedural memory (in how the brain encodes and also, one could argue, in how performance is assessed and measured). Therefore it is possible that the effect of odor, in this example, could differ for procedural and declarative learning and thus motivate unique examinations.
- **Consider whether there are two possible changes of your treatment** (or intervention or experimental variable) and these are mutually exclusive (i.e., one could happen even if the other doesn't). This provides another angle. For example, using the overarching aim

above, I might consider Aim 1 *"To examine whether odor cues associated with learning enhance performance of older adults when re-presented during subsequent recall"* and Aim 2 may be *"To examine whether the neural representation of the memory is altered when an odor cue present at learning is re-presented during subsequent recall (using fMRI techniques)."* In this case, I must presume that it is possible for behavior to change (Aim 1) even if the underlying neural representation doesn't (Aim 2) and vice versa (see below "House of cards").

> Now write out a **statement of your specific aims**. For example, *"To achieve this overarching objective, the specific aims of this proposal are:*
> *Aim 1. To examine whether odor cues associated with learning of a declarative memory task enhance performance of older adults when odors are re-presented during subsequent recall.*
> *Aim 2. To examine whether odor cues associated with learning of a procedural memory task enhance performance of older adults when odors are re-presented during subsequent recall."*

Now check your aims for these mistakes
- ***Disconnected.*** Sometimes students can generate two good ideas (aims) but they lack the thread to tie them together to allow for a single, overarching aim. It's not uncommon for reviewers to complain the aims were disconnected or disjointed. Being disconnected also makes it hard to write the Introduction/Background and you don't have a lot of space to work with. So you might need to rework them before you move on.
- ***House of cards.*** Does Aim 2 require that your hypothesis to Aim 1 prove correct? What if your hypothesis for Aim 1 is not supported (or the study fails), is Aim 2 still necessary/interesting? If these are the case, then you have created a house of cards and your work is not likely to be supported. [Note that a sort of choose-your-own-adventure (if this happens I will do this and then if this happens I will do this but if this alternative happens, I will do this instead) are often approved for a dissertation proposal but not accepted by grant reviewers very favorably (although there are exceptions).]
- ***Null result.*** We will get to formulating your hypothesis next, but it will be critical that your aim does not require you to prove the null hypothesis. In other words to say that you will *not* find something significant will set you up for statistical failure. For example, we could not have an aim that states "To examine whether odor cues associated with learning **do not** enhance performance of older

adults when re-represented during subsequent recall."

While the aim defines a scope of something you plan to do, the end goal of every aim is to test a hypothesis. This is where we can reflect back on a basic science or methods course to define a hypothesis.

This section may seem elementary, but even successful researchers have made mistakes in providing testable accurate hypotheses so it is worth revisiting.

What is a hypothesis? You may think this is too basic for a graduate-level handbook. We all know that **to hypothesize** is to *surmise that something may be true*. However, you might hypothesize that *cows can sense when it will rain* but that would not be a good hypothesis. Why not? Because a good hypothesis provides an explanation, and is clear, testable and measurable. Let's see how it holds up.

I hypothesize that cows can sense when it will rain.		
Explains what you expect to be true.	✅	I expect that cows can sense when it will rain.
Is clear	❌	What sense might the cow be using?
Is testable	?	Without knowing what sense the cow is using we don't know whether this is testable.
Is measurable	❌	How would we measure that it is indeed predicting the rain?

So consider another hypothesis.

I hypothesize that cows lay down before it rains.		
Explains what you expect to be true.	✅	I expect that cows lay down before it rains
Is clear	✅	Seems pretty straightforward
Is testable	❌	'Before it rains' is an infinite window and assuming cows lay down at any point in their lives, your hypothesis will always prove correct. But can you measure 'laying down' for the lifespan of a cow? And is that indeed the intent of the hypothesis?
Is measurable	✅	One could manually note when a cow lays down. Or perhaps there are body sensors for cows that would provide this measurement.

Ok, ok, you're thinking this is a mistake that only those making silly hypotheses about cows and rain make. But it's actually not. It is an error found even in faculty grants. Let's consider more realistic example that might go with Aim 1 described above.

I hypothesize that odor cues associated with learning will enhance performance of older adults when re-presented during subsequent recall.		
Explains what you expect to be true.	✅	This seems to explain what I expect to be true.
Is clear	❌	"Will enhance performance" is unclear. Enhance relative to what? Relative to young adults? Relative to recall in the absence of odor cues?
Is testable	✅	Assuming the clarity issue can be addressed, it seems reasonable that you can test memory for items learned with an odor cue (and perhaps test with and without re-presenting the odor cue)
Is measurable	?	Presumably. However, it may depend on what is learned. Typically a hypothesis statement does not stand alone so the task may be described in a prior sentence.

So let's end with a 'right' example of a hypothesis:

I hypothesize that odor cues associated with learning of face-name pairs will enhance performance of older adults when re-presented during subsequent recall relative to recall in the absence of the odor cues.		
Explains what you expect to be true.	✅	
Is clear	✅	Now it is very clear what conditions are being compared – a condition with and without odors at recall.
Is testable	✅	
Is measurable	✅	We now know that face-name pairs are learned and this can be assumed to be a measurable behavior.

Now we have a good hypothesis. Note that it's a very long sentence. We'll return to this later.

CHAPTER THREE

Testing Your Scope

While you could begin by writing your Methods section, you may get too hung up in the writing and the words to see the scientific details. So, I suggest starting with a sketch. What your sketch will look like really depends on your research area, but we can attempt to generalize.

<u>Location.</u> Where will your research take place? For most people, this will be straightforward – it will take place in the lab that you belong to. But not always. Will you need to collect data on a rare bird in the Amazon? Or use equipment not available to you on your campus?

<u>Source of data.</u> Where is your data coming from? Are you studying humans, mice, or tissue samples? How many will you need? Pencil a number in. Perhaps you are a chemist and will mix reagents. What are these? How much will you need? Can you justify how many you will need? Where will you get them? What gives you confidence that you can get this resource?

<u>Measurement and data.</u> What is your measure of interest? What is the data you are collecting from this source? Are you giving paper questionnaires to a humans? Are you measuring learning in mice? Are you measuring the presence of a molecule in a tissue sample? You have something you are measuring in one form or another. Perhaps you are doing a qualitative study of living conditions in some population; your data is the responses of these people.

<u>Expected outcome.</u> From your hypotheses, you have an expected outcome. But what does this look like in terms of your data? You

should have predictions that something measurable will happen. What is this measure in your data? Will your data answer this hypothesis?

You may hit a few snags in your approach as you sketched it. Regardless, it could save precious time later to check now that you have a viable plan. Here are some 'checks' on your responses.

Location: If any part of your project will take place somewhere other than your campus, will you be able to access it? It may require an additional (substantial?) budget. Does the award you are applying for cover travel? Some awards will not allow funds to be spent internationally (and some foundations may limit spending to certain regions as well).

Source of data. Most importantly, consider whether you can access this resource. For example, you may be studying older adults, as in the example, but can you recruit enough of them fast enough? If you live in a non-urban area, reviewers are apt to have concerns about this. Perhaps you plan to study a mouse disease model. Is this accessible and affordable? Carry out a similar line of questioning with your source.

Measurement and Data. Will you have access to the tools you need to do the measurement? Do you have experience or access to training with the tools you need?

Expected Outcome. The most important 'check' here is that the data you're planning to collect actually addresses the hypothesis you formulated.

Once you know you have a viable study, let's check that it fits the timeline of the award. It's easy to be overly ambitious. So, think it through in great detail. Depending on the grant you are applying for, a detailed timeline can be useful for convincing reviewers that not only it will fit into the time allotted but also that you have thought it through thoroughly.

The example below illustrates a carefully considered timeline. It takes into account the length of the funding period (in this case, 3 years) and the unique timeline of each experiment, and additional time needed to analyze data and prepare the manuscript.

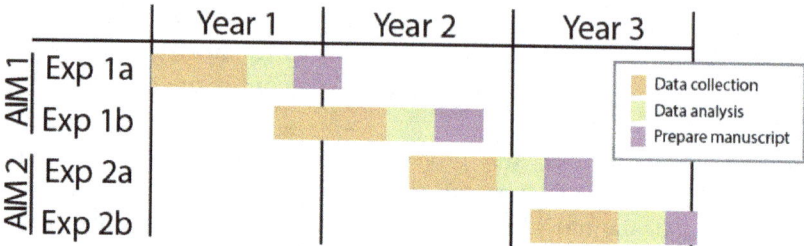

Be sure to consider a reasonable time for aspects like analysis and writing (as illustrated above which is likely overly ambitious!). In some cases (as illustrated above), analyses may take place in a few months' time. But for some forms of data, analysis may take longer than data collection. Perhaps your type of science has different or additional categories – like focus groups, preparing transcripts, or designing and testing a molecule. Break the research into reasonable key steps that fit your work..

You probably identified any issues with timing as you drew out your timeline. Think carefully through how long it will take, reasonably, to collect your data. Reviewers understand the pace of data collection for a graduate student project may be less than that of their PI's project that may be staffed with more people or run by post-doctoral fellows particularly since you will be including training components in your proposal as well. But it is a fine line! They want you to be ambitious but not too ambitious.

Review your timeline sketch and consider the following:
- Some overlap of phases is reasonable and, in fact, recommended. For example, I can only collect data at certain times of day so I will analyze or write Experiment 1 while I am collecting data for Experiment 2.

- If your grant has a significant training component, for instance, you will travel somewhere to learn a technique first, be certain to account for that in your timeline.
- Do you have justification for your estimates for the length of time it will take to collect the data? Have you or a peer in your lab done something similar to make you confident in this guess? The pace of the data collection phase is often under-estimated. If you have not yet done studies like those that you propose, be sure to consult with your advisor and peers in your lab to come up with an accurate estimate.

CHAPTER FOUR

Some Advice on Good Writing

First, it is important to give yourself time to write. The NIH F31 is long (it may end up to be 60+ pages when all is said and done!). While the NSF GRFP is short (only 5 pages of writing!), it can be time-consuming to develop your thoughts. Also, expect to go through many iterations of your documents. Some trainees joke about the number of drafts; ResearchStrategy_v32.doc may sound like a joke until you've gone through the process yourself!

Some tips to make your writing as efficient and productive as possible:
- Try to set aside a little time every day. Many suggest that 1 hour each day is a good goal. During this 1-hour, turn off email and all distractions. It is a solid hour of writing, spewing words onto your paper. Clean up later.
- It is also good to schedule this time early in the day, even first thing, when you are 'fresh'.
- Write free from distractions - turn off email, leave your phone in a different room, choose a location that is quiet and comfortable.

Most importantly, look ahead at how much time you have, and create a timeline. Find a peer who might (or not) be writing a fellowship application too and hold each other accountable for making the deadlines on your timeline. Here are an example timelines for different awards.

* * *

NIH F31 Timeline

	April 1	April 15	May 1	May 15	June 1	June 15	July 1	July 15	Aug 1	Aug 15	Sept 1	Sept 15	Oct 1	Oct 15	Nov 1	Nov 15
Getting Started, Testing Scope, Advice	X	X														
Specific Aims		X	X	X	X	X	X	X	X	X	X	X	X	X	X	X
Selection of Sponsor & Institution		X	X	X	X	X	X	X	X	X	X	X	X	X	X	X
Research Strategy			X	X	X	X	X	X	X	X	X	X	X	X	X	X
Training in Responsible Conduct				X	X	X	X	X	X	X	X	X	X	X	X	X
Biosketch (fellow)					X	X	X	X	X	X	X	X	X	X	X	X
Background & Goals						X	X	X	X	X	X	X	X	X	X	X
Facilities & Other Resources, Equipment							X	X	X	X	X	X	X	X	X	X
Reference Letters, Letters of Support, & Letter of Eligibility*								X	X	X	X	X	X	X	X	X
Sponsor Requests (Sponsor/co-Sponsor Statement, Biosketch)									X	X	X	X	X	X	X	X
Institutional Environment and Commitment to Training										X	X	X	X	X	X	X
Respective Contributions											X	X	X	X	X	X
Project Narrative, Project Summary/Abstract												X	X	X	X	X
Human Subjects and/or Vertebrate Animals													X	X	X	X
Resource Sharing Plan, Cover letter														X	X	X
Bibliography/References Cited						X	X	X	X	X	X	X	X	X	X	X

NSF GRFP Timeline

	April 1	April 15	May 1	May 15	June 1	June 15	July 1	July 15	Aug 1	Aug 15	Sept 1	Sept 15	Oct 1
Getting Started, Testing Scope, Advice	X	X											
Objectives		X	X	X	X	X	X	X	X	X	X	X	X
Personal, relevant background, future goals			X	X	X	X	X	X	X	X	X	X	X
Research Plan					X	X	X	X	X	X	X	X	X
Applicant Data						X	X	X	X	X	X	X	X
Reference Letters								X	X	X	X	X	X

X = drafted; X = completed; ☐ = first draft completed; ☐ = revised draft completed

To write a good proposal, you need to write well. Writing styles vary. However a feature of scientific writing, particularly for grants with word limits, is efficiency with words - use as few words as possible to make your point. But also avoid words and errors that might bug your reviews. Here are some pointers:

1. **Avoid metadiscourse.** Metadiscourse is the tendency to write

things with far more words that you need.[1]
   a. Good: Sleep facilitates memory consolidation
   b. Bad: Sleep <u>has been shown to</u> facilitate memory consolidation

   a. Good: Between 6-15 months, infants transition to two naps per day.
   b. Bad: Between the ages of 6-15 months, studies have found that infants transition to two naps per day.

   When you edit your own documents and particularly as you edit others (it is always easier to see in someone else's writing!), look for metadiscourse. As you can see, this writing is not "bad" per se, but it is not concise and when embedded in an already long sentence, can make the writing hard to read. Also, you will find that **words are precious in grant writing!** You are given limited space so choose your words carefully!

2. **Avoid acronyms.** This is a common mistake: if you use acronyms, be sure to introduce it on first use and then use it consistently for the rest of the document. Also, if you do introduce an acronym in the Specific Aims, you will need to redefine it in the Research Strategy. However, it is highly recommended to avoid acronyms and abbreviations in short sections such as the Specific Aims. As a general rule, use acronyms that are very familiar to even people outside your narrow field. It is recommend that you limit your scientific section (NIH Research Strategy or NSF Research Statement) to no more than three acronyms.

3. **Use logical flow.** The reader should be able to follow your train of thought or guess what your outline was. Use an outline to develop your writing and see if a friend can guess your outline when you are done with a section.

4. **Read, read, and read again.** When someone reads your

---

[1] I credit this concept of metadiscourse in writing to Dr. Peter Sterling. I also credit my lab staff over the years for tracking my pet peeves in writing.

proposal, it is easy for them to be distracted by editorial errors. So don't waste the time of your volunteer editors with a bad draft, particularly your advisor or co-sponsor. Unfortunately, some advisers only read two or three drafts at most. Don't waste one of these on editorial clean-up duty that you could do yourself.

6. **Write for a broad audience.** If your roommate is in physics or something distantly related to yours, ask them to read it. This is particularly true for an NSF application.

7. **Avoid overly profound statements.**
    a. Good: Sleep benefits memory.
    b. Bad: The purpose of sleep has been a mystery for thousands of years.

8. **Use "I" or "my" very sparingly.** It's okay if you are referring to a study you have done in your lab ("Work in our lab has shown..."). Also, for fellowships, take ownership of ideas and plans so **if you use a pronoun, use "I" rather than "we"**. These pronouns will be essential in some sections (Background and Goals) but not in others (Research Strategy).

9. **Avoid casual sentences** (in other words, be careful not to write like you talk, with colloquialisms).
    a. Good: Participants are required to have normal or corrected-to-normal vision to be eligible.
    b. Bad: Participants have to have good vision

10. **Use author names in the text sparingly** (only when it is particularly relevant, such as pointing to conflicting research). For example, it is not necessary to say "Jones and colleagues (2019) showed..." unless it is particularly relevant that it was Jones. Likewise, no need to say "In a study conducted at the University of Tokyo...." unless it is relevant to your point that this work was done in Tokyo.

11. **Avoid loaded terms.** For instance, "modulate", "interact" are examples of terms that have technical definitions. Unless you

are using these according to their statistical purpose, do not use these terms.

12. **Keep verb tense consistent.** Chose a tense and stick to it.

13. **Keep the number consistent.**
    a. Good: After the participant completes the informed consent process, the participant will be given a set of questionnaires...
    b. Bad: After participants complete the informed consent process, the participant will be given a set of questionnaires...

14. **Do not use contractions in a grant proposal.** While contractions seem to have gained acceptance in some types of formal writing (including this book!), they should be left out of a grant proposal (and scientific papers in the humble opinion of this author).

15. **Words to avoid:**
    - revealed: this is often used incorrectly and implies some sort of magic.
    - displayed: use this carefully.
    - upon: usually you mean 'on'
    - within: usually you mean 'in'

A note on citations. Although this is often a matter of debate, I recommend using numbered, in-text citations, in particular, superscript numbered citations. Reasons for this are: (1) these take the least amount of space and space is critical and (2) a lot of cited names makes citations long which makes the text hard to read. That your citations must visually give a shout out to elders in the field is an old wives' tale. Reviewers may want to have recognition of certain findings but this does not have to be evident by having names in citations (the reference list will do). The case is also made that you should write out names in citations to draw attention to when you cite yourself. But this is what your biosketch is for. Instead use in-text cues such as: "In a previous study in our lab..."

* * *

Importantly, it is HIGHLY recommended that you use a reference manager. Start now, do not wait until later (although you can thank me then). Mendeley (www.mendeley.com) and Zotero (zotero.org), for instance, are free reference managers and will help you manage your references both for this proposal as well as your papers and dissertation in the future!

FORMATTING IS IMPORTANT!

I say that with caps to get your attention. This shows how formatting can be used to enhance your writing. There are a few reasons I will highly emphasize formatting:

1) Some formatting errors will get your grant returned without review. It is important to follow the funder-issued formatting (length, size margins, font size) as violating these can get your grant returned without review.

2) Formatting errors can waste time later. Often formatting errors are picked up before you submit - by a mentor or perhaps a grants office. However, getting these things right in the first place will save your time for other important final edits that you will want to focus on in the end.

3) Inconsistent formatting is obvious to reviewers and can indicate carelessness. Although proposals are composed of multiple sections and each is created and uploaded separately, a grant reviewer will see all the sections <u>as one continuous document</u>! Inconsistent formatting from one section to the next is quite evident. Consistent formatting makes a grant appear more meticulous and professional. Also, sections do not automatically get labels so a reviewer can get lost if you do not consistently insert a header (name of section).

4) Formatting can be used to highlight key points you want to draw the reviewer's attention to. Throughout this book you will see that I use bold, italics, and underling to highlight points. Apply this method to your grant writing to make your train of thought and key points clear.

5) Proper formatting can often save you space. Setting your margins first will help you realize the space you are working in and it is important to fill it. We will return to this point later.

Box 2. Overview of requirements for the NIH F31

Specific Qualifications:
- Must be a U.S. Citizen or permanent resident
- Enrolled in a PhD program in the biomedical, behavioral, or clinical sciences
- In the dissertation research stage
- Evidence of "high academic performance in the sciences, and commitment to a career as an independent scientist"

Overview of Components:
- Proposal Summary/Abstract (limited to 30 lines of text)
- Project Narrative (limited to 3 sentences)
- References Cited
- Facilities and Other Resources
- Equipment
- *If applying for Diversity Fellowship:* Description of Candidate's Contribution to Program Goals
- Biographical Sketch
- Applicant's Background and Goals (limited to 6 pages)
    - Research experience
    - Training goals and objectives
    - Activities planned under the award
- Specific Aims
- Research Strategy (limited to 6 pages)
    - Significance
    - Innovation - optional
    - Approach
- Respective Contributions (limited to 1 page)
- Selection of Sponsor and Institution (limited to 1 page)
- Training in Responsible Conduct of Research (limited to 1 page)
- Sponsor and Co-sponsor statements (limited to 6 pages)
- Letters of Support from Collaborators, Consultants, and Contributors (limited to 6 pages)
- Institutional Environment and Commitment to Training (limited to 2 pages)
- Resource Sharing Plan
- Human Subjects (if relevant)
- Vertebrate Animals (if relevant)

Getting started.

The list above provides the components in the order that the proposal will be compiled. However, this book will have you develop these sections in a different order chosen to manage your time best.

Formatting.

Before starting any section, format your document! Although you will develop each section separately, they will be compiled to one (long) document for your reviewer. Consistent formatting across your sections makes the proposal much more crisp and professional. So, while you may not need tight formatting for space reasons on some of the supporting documents, the need for tight formatting for space in the Research Strategy is why you should apply that spacing to supporting documents as well.

Format as follows:
   Set page margins too .5 inches
   Set font to 11-point
   Set font to Arial (or similar) font
   Set line spacing to single spacing
   Preferred: Set line spacing to 4 to 6 pt after each paragraph. To do this, go to: 'line spacing options'-> under 'spacing'-> set 'After' to 4 pt (or 6 if you have enough room).

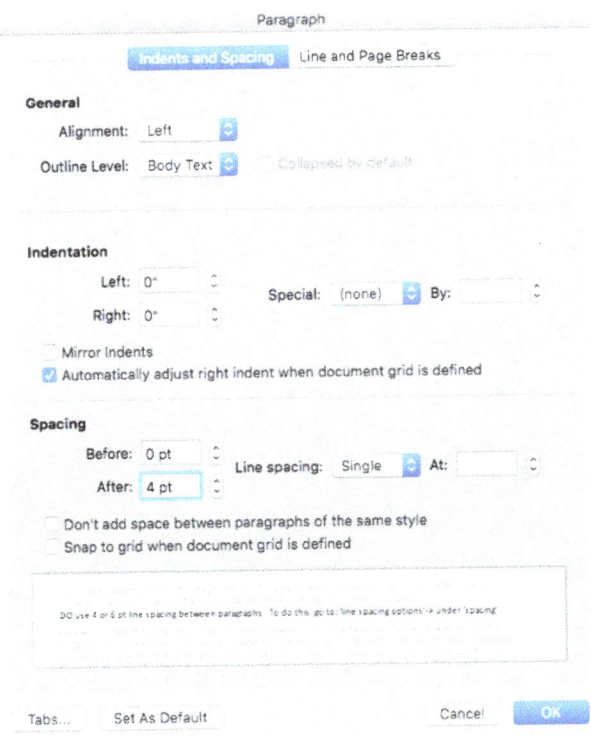

Note: This spacing between paragraphs will delineate paragraphs sufficiently so <u>do not</u> indent paragraphs.

Additionally, pay very careful attention to your headers on each document. For instance, I recommend every document have the section title on the first line, left aligned, all caps, in bold. As in:

SPECIFIC AIMS

CHAPTER FIVE

Specific Aims

The Specific Aims page is like the abstract for your research proposal (although you will do an even shorter version, called the Project Summary, later). This one-page document should summarize your research concisely. Because it provides a standard overview, it is a very important document for reviewers. It often sets the stage for the reviewer - they tend to read it early as they wrap their mind around what you plan to do. It also serves as a concise summary for those not assigned to review your full proposal.

It is also helpful to develop the specific aims first as it forces you to really hash out your research idea and justification without getting all the way into the longer Research Strategy.

> **Format requirements:**
> 1 page limit
> Standard format (.5 inch margins, 11-point Arial font)

Always start with formatting your document so you know how much space you are working with (you will need it all). In addition to setting up your page, put the section header on the first line:

SPECIFIC AIMS

From here, you will write:
 An introductory paragraph to set the stage

A paragraph specifying your objectives
Aims paragraphs
Significance paragraph

NOTE: This format for the Specific Aims is derived from that of the Grant Application Writer's Workshop - NIH version by Stephen Russell and David Morrison and developed here to apply for a fellowship. See the Grant Application Writer's Workshop guides for more information.

Paragraph 1: Set the stage

The first paragraph is important to 'sell' your work right away. For NIH it is important to sell your work as being fitting to your specific funding agency. Funding for the National Institutes of Health requires health-relevant research. Their mission statement is:

...to seek fundamental knowledge about the nature and behavior of living systems and the application of that knowledge to enhance health, lengthen life, and reduce illness and disability.

But they also state:

*"the NIH provides the leadership and direction to programs designed to improve the health of **the Nation**."*

Although in many cases that may seem obvious, it is important to point out why your health-related question is important to health with a U.S.-emphasis. For example:

"Depression is one of the most common comorbid conditions in Alzheimer's disease and related dementias and is also a risk factor for developing Alzheimer's disease."

"Alzheimer's disease presents a significant cost, including the economic burden of healthcare and a burden on caregivers."

*"Sleep deprivation has a significant health and economic impact, with reduced productivity and increased risk of mortality **costing the U.S**. over $411 billon per year."*

In this first paragraph, start with your statement of relevance that is reasonably broad. After that, provide just a few sentences of critical background. End the first paragraph with a statement of the gap.

Paragraph two: Define your goals

The second paragraph should place your research plan in the context of your career objective. This is, after all, a fellowship to fund you developing your training for your career and the research is just one piece of that. This statement should align with your Background and Goals section (we will get to that later).

For example, this second paragraph could start with a sentence like:

"My long-term goal is to be in independent investigator at an R1 research institution with a research focus on sleep in Alzheimer's disease."

Follow this with a sentence stating your specific research goal as part of this fellowship which should also align well and seem like a first step towards that long-term goal. Such as:

"The specific objective of this proposed research, a critical step towards my career goal, is to examine whether sleep is impaired in individuals with mild cognitive impairment."

Now state your overarching hypothesis.

Depending on your aims, you might end this paragraph with a summary of your approach/methods. However, if they will differ by aim, you can save that for your aims paragraph.

Aims paragraphs

You will have one paragraph for each of your aims. Start each paragraph by stating your aim.

If you have different methods for each aim, then use a sentence or two to give the reviewer some sense of what you plan to do.

Finally, state your working hypothesis.

Significance paragraph

The final paragraph of Specific Aims should state the significance of your work. I recommend stating two forms of significance: the scientific significance of your work (theoretical/translational) and the significance to your career development.

SPECIFIC AIMS

Sleep is important for physical health and brain plasticity in young adults. Yet, little is known about the function of sleep in children. *[interest-grabbing statement]* Between 3-5 yrs, sleep transitions from biphasic (mid-day naps and overnight sleep) to monophasic (single overnight sleep bout). In spite of this transition in circadian and homeostatic sleep processes amid a sensitive learning period, whether sleep serves a cognitive function at an early age is unknown. *[current knowledge]* There is an urgent need to understand the function of sleep, and particularly naps, in children. Lacking evidence of academic benefits from mid-day sleep, increasing financial and curriculum demands on preschools have led to shortening and elimination of nap opportunities. *[gap in knowledge/unmet need]*

The long-term goal of my research is to understand sleep function on cognitive, motor, and emotional development, and the implications of sleep function for enhancing well-being. *[your long-term research goal]* The **specific objective** of the proposed research is to examine the function of sleep on multiple forms of learning in young children (3-5 yrs) using naps as a model. *[objective of THIS research]* The <u>central hypothesis</u> of this proposal is that naps benefits recent memories and subsequent learning for preschool children. *[overarching hypothesis]* In support of this hypothesis, our preliminary data in 35 children illustrates that when learning is followed by a nap, recall is <u>protected or enhanced</u>, whereas when the same children spend an equivalent interval in quiet wake, a 10-14% decrement in recall is observed. *[what the hypothesis is based on]* The <u>rationale</u> for the proposed research is that, by demonstrating that sleep benefits learning in healthy children, sleep can be considered a novel target for enhancing education, particularly for children with developmental disorders and psychopathologies. Thus, demonstrating the benefits of sleep on learning for healthy children across a range of tasks, will lay the groundwork for subsequent studies examining whether sleep promotion may enhance well-being in disadvantaged populations (e.g., at-risk, learning delayed). *[rationale]*

Aim 1 is to determine the function of mid-day naps on <u>prior</u> learning across task domains for preschool children. Sleep enhances memory consolidation in young adults. To assess whether memories are likewise modulated over a nap for preschool children, *[method statement]* children will learn a task and recall will be assessed following a mid-day nap interval and an equivalent interval spent awake. Given that sleep is not homogeneous – with distinct sleep stages associated with declarative and procedural learning – we will probe these independently. Based on preliminary data, my working hypothesis is that mid-day naps function to enhance and protect declarative and procedural memories: Recall will be greater following a mid-day nap relative to recall when the same children (within-subject design) spend the mid-day rest opportunity in quiet wake. Importantly, I expect the benefit of the nap will remain when recall is probed the following day, when sleep pressure is equated for the nap- and wake-promoted conditions.

Aim 2 is to determine the function of mid-day naps on <u>subsequent</u> learning across task domains for preschool children. Sleep enhances encoding of new memories in young adults. To assess whether naps serve this function for preschool children, following a mid-day nap or equivalent interval spent awake, children will be taught a declarative or procedural learning task. Recall will be assessed following the nap/wake interval and again the subsequent day when sleep pressure is equated for the nap- and wake-promoted conditions. Based on preliminary data, my working hypothesis is that mid-day naps function to enhance subsequent learning across task domains, a benefit evidenced in the short (immediate recall) and long term (delayed recall). *[specific aims paragraphs]*

IMPACT: The outcomes of this proposal have translational **significance**. Naps are an innovative, low-cost, and low-burden intervention that may benefit learning and enhance retention, leading to improved school readiness and subsequent health benefits. Naps could be particularly beneficial to children in underserved populations and children with developmental delays and psychopathologies that impair learning. *[significance]*

Through the course of this research, I will gain critical research experience, including learning sleep measurement techniques (polysomnography and actigraphy) and apply analysis approaches learned through coursework. This research will also provide multiple opportunities for presenting interim summaries at research conferences and for manuscript publication. *[career impact]*

CHAPTER SIX

Selection of Sponsor and Institution

It is important to convince the reviewers that you are at the right place for both you and your research interest. This may not be the way you thought of your choice of graduate schools but it is important that you now recognize that you are a great match with both the Institution (your university and graduate program in particular) as well as a great match with mentor.

Your grant will be scored in part on whether your mentor (sponsor*) is a good fit for you and whether you are in the right place to meet your objectives. In this section you will convince the reviewers of this by emphasizing their fit for your research style, research objectives, and preparing for your career ahead. It requires salesmanship.

*An important point of clarification. In an NIH F31 (or F32) application, you, the graduate student (or post-doc), are considered the **principal investigator** (PI) or candidate or applicant. Your mentor is considered the **sponsor**.

> **Format requirements:**
> 1 page limit
> Standard format (.5 inch margins, 11-point Arial font)

1. Start by formatting your page to the requirements above. Next, put the section title at the top:
   SELECTION OF SPONSOR AND INSTITUTION

2. Now, you will create 2 or 3 headers:

Selection of Sponsor

Selection of Co-sponsor (optional)

Selection of Institution

3. Now it's time to start writing.

- **Selection of Sponsor:** In this section, start by describing what drew you to work in the lab of your PI? What is their experience (or potential) as a mentor?* What is your PI's expertise and can you reference accomplishments (awards they have won, prominent research findings, or unique tool or technique they are known for). After describing these practicalities of matching research-wise, turn to selling this person as a mentor and how you are a good fit for this mentoring style. It is important that this paragraph emphasize their scientific perspectives and mentoring style are ideal for what you want to do. Also, emphasize their skills that made you select them rather than your experiences together since you started (that will go elsewhere!).

- **Selection of Co-sponsor:** First, it is worth saying a few things about whether you have or should have a co-sponsor. A co-sponsor is not required but can be useful in many cases. The co-sponsor's role is to provide additional training and mentoring, filling gaps of the sponsor. For instance, if your research will combine two different areas of expertise, or if you will bring a new technique to the work of your mentor's lab, these would be reasons to add a co-sponsor.

    If your mentor is junior, meaning they have only recently started their independent research position and may not have a significant history of grad student mentoring. This would be a case where having a co-sponsor with a record of grad student mentoring (as well as research relevance) would be strong. This is not required and NIH reminds reviewers that junior investigators can be sponsors on fellowships. Nonetheless, the gap that a junior professor might have in their ability to mentor you is still a potential target for reviewers that you want to preempt.

If your mentor has not had recent research funding, a co-sponsor with a strong funding history may help your application (although you should consider where funding for the work itself will come from). MOST IMPORTANTLY, reviewers will want to know - where will the money come from to do what you propose? The fellowship will fund you. What funding source will cover the cost of your study and workshops or travel you propose? Sometimes a co-sponsor is willing to take this on (or be your financial cushion).

So the content of this section should clarify why you chose this person as your co-sponsor. Similar to the Selection of Sponsor, explain the co-sponsor's expertise as well as mentoring fit.

It is also possible to have two co-sponsors. This may seem ideal to balance strengths and weaknesses. However, it is not recommended to have more than one co-sponsor. There are a few reasons for this. First, the Sponsor/Co-Sponsor Statement (see Chapter 13) will have 6-pages that must be shared by your sponsor and all co-sponsors. It is very hard to split this between 2 sponsors let alone more. Second, it may begin to look unfeasible, or minimize the input you will get from each. You would also have to balance the time that it would take to have too many people to report to. **It is great to have a team,** but think of other roles alternative people might play (e.g., collaborator? consultant?).

- **Selection of Institution:** Start by describing your selection of the specific graduate training program you are in. What are some assets of this program? Is your interest or specialization something that is represented elsewhere in your program (e.g., multiple professors in your graduate program study memory from different angles)?

Then, in a second paragraph, speak to the institution more broadly. Does your program affiliate with other programs? Does your institution have specific resources or centers or core facilities (which you will describe in detail elsewhere but you might refer to here as a reason it is a good fit)? Is it situated in an ideal location to access other opportunities?

*If **you work with a new faculty member**, they may not have demonstrated success as a mentor as no trainees have yet graduated from the lab. Consider ways that you can potentially counter this concern. First, has the sponsor mentored other trainees, such as undergraduates, or played a significant role in graduate student training while a post-doc? Describe these mentoring experiences. Perhaps more importantly, emphasize the mentoring experience of the co-sponsor. The co-sponsor should be described as filling that gap. Finally, consider the situation in which **both the sponsor and co-sponsor are junior faculty** members without a significant history of mentoring success. In this case, it is very important to consider the role of senior mentors in your training (and indicate this in the Respective Contributions section). However, for this reason, I do not recommend having two junior faculty sponsors.

CHAPTER SEVEN
Research Strategy

The Research Strategy is composed of Significance, Innovation (optional), and Approach. Although these are the titles of the components, there is a lot that you will need to convey in these sections:
- Necessary background for the reviewer to understand your area and the problem (and pay due respect to other tightly related work in the field)
- Research methods
- Analysis plan
- Potential limitations and pitfalls
- Evidence of feasibility
- Detailed research timeline

So it will be important to use this space wisely. The description that follows will help guide you to constructing these sections with the necessary goals in mind.

> **Format requirements:**
> 6 page limit
> Standard format (.5 inch margins, 11-point Arial font)

There are many points to get across in a small amount of space. So it is helpful to use formatting and bullet points, **or what I think of as stepping stones,** to make your work conceptually clear. Like taking the most direct route through the garden, the stepping stones could tell the whole story but with no details. But just like you could step off any stepping stone and look at the flowers more closely, the text after each stepping stone, the rest of the paragraph, provides the details of the story.

> **Box 3. Stepping stones and formatting in writing a research strategy**
>
> Stepping stones are summary sentences that start each paragraph and could stand alone for someone browsing your application. Formatting tricks (like **bold** text and underlining) can also be used to help your reader follow you conceptually or for an unassigned reviewer to browse it and get the gist of what you are saying.
>
> Here is an example of a passage from a Research Strategy:
>
> ---
> **RESEARCH STRATEGY**
> **1. Significance**
> **Sleep supports plasticity.** This finding has led to the recognition of deficits in psychiatric disorders (e.g., schizophrenia[1,2]) and sleep disorders (e.g., sleep apnea[3,4]) that reflect impairments in sleep-dependent plasticity. Thus, understanding sleep's role in plasticity across the lifespan holds promise for recognizing and treating sleep and psychiatric disorders. <u>Understanding this process during development is key to early intervention.</u>
>
> **Sleep benefits learning in infancy and early childhood.** A number of studies have provided evidence that infant memory is better following an interval of sleep (a daytime nap) relative to an equivalent interval spent awake for declarative[5-8] and procedural memory[9] as well as generalization.[6,10-14] For example, napping facilitates declarative memory in 6- and 12-month-olds.[5] Using a deferred imitation task, infants saw an experimenter demonstrate an action with a novel object (an unknown toy). Infants' imitation of the experimenter's actions (recall) was probed both 4 and 24 hours later. Infants who napped for ≥30 mins showed greater recall of the actions following a 4-hour delay compared to a control group who did not observe the actions. Notably, infants who remained awake or took brief naps (<30 mins) performed similar to the control group. Sleep has been likewise shown to benefit learning in early childhood (3-5 yrs),[15-18] adolescents,[19,20] and young adults.[21,22]
>
> **Sleep benefits are thought to reflect memory consolidation.** <u>Memory consolidation</u> is a process by which new memories are stabilized and become more resistant to interference.[23] Unlike <u>memory encoding</u> and <u>memory recall</u>, memory consolidation takes place off-line and without awareness.[24] Although memory consolidation can take place over wake and sleep, sleep provides an ideal environment for memory processing, and thus consolidation takes place preferentially during sleep.
>
> **Memory consolidation can take place at the synaptic and systems levels.**[25] At the <u>synaptic level</u>, synaptic downscaling of some synapses while other parts of the engram go through synaptic upscaling improves memory. Both synaptic upscaling and downscaling are enhanced by sleep,[26] particularly slow wave sleep.[27] At the <u>systems level</u>, memories that are initially hippocampal-reliant are transferred (or copied) to the neocortex, where they are integrated with related memories.[28] Memories are reactivated in hippocampal CA3 subfield, and these hippocampal
>
>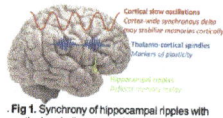
>
> **Fig 1.** Synchrony of hippocampal ripples with cortical spindles and slow oscillations is
>
> ---
>
> A few things to note:
> - Page formatting: The headers at the top of the page as recommended.
> - "Stepping stones": These are the bolded first sentences of each paragraph. You could follow the outline, or train of thought of the writer, just by looking at the bold statements. You can find things easier with these as well.
> - Other formatting tricks: Underlined words stand out and draw emphasis to key statements or words that need to be defined.
> - Use of space: Break the page up with figures and lines (small spacing!) between paragraphs.

Stepping stones can also make the writing process easier. Start by creating the bullet points that you want to get across in this section – almost as if you were creating a Powerpoint slide of what you want to say; the bullets tell the key point but your speech would provide more detail for each. Importantly, for stepping stones to work, the statements must be short and concise (again, like you might have for a bullet point in a presentation - you don't say all the words in one statement, just the gist/summary of the point).

* * *

In general, your points may roughly follow this template:
- **Statement setting your work in the context of human health significance. Paragraph should end with pointing to gap.**
- **Background: statement of important fact #1**
- **Background: statement of important fact #2 (optional but likely needed)**
- **Background: statement of important fact #3 (optional)**
- **Statement of your overarching objective**
- **The proposed work has theoretical significance**
- **The proposed work has translational significance**

Start by writing these bullet points. **Bold** the text.

Next, following each bullet, write a brief paragraph providing the support for your bolded statement.

Note that, under these statements, you may have actual bullet points. For instance, there may be 3 areas of translational significance. Individuate each one with a bullet.

Also, as a general rule, it is advised to have a figure or table on most pages of your proposal. A figure in this section may illustrate a process or idea or perhaps data from a key prior study that is essential to motivating your planned research.

Here is an example of what that might look like:

RESEARCH STRATEGY
1. Significance
Understanding age-related changes in sleep and memory in healthy older adults provides a critical foundation for Alzheimer's disease research. An overarching theme of the 2015 Alzheimer's Disease Research Summit was the need to understand all aspects of healthy brain aging and cognitive resilience to inform Alzheimer's disease prevention.[1] This proposal is significant in this regard. Sleep is disrupted in Alzheimer's disease.[2,3] Thus, changes in sleep may underlie some changes in cognitive function, particularly early in the course of Alzheimer's disease (i.e., amnestic mild cognitive impairment).[4] However, this prediction assumes a cognitive function of sleep in healthy aging, which is not yet fully understood.

The overarching objective for our research is to understand the contribution of sleep to cognitive function in aging. Sleep is impaired even in normal/healthy aging. Most significant are increases in nighttime awakenings, time of sleep onset, and daytime sleepiness. Total sleep time and sleep efficiency decrease,[4] and slow wave sleep (SWS) and rapid eye movement (REM) sleep are reduced.[5,6] Although non-REM stage 2 (nREM2) is preserved or even increased,[5] the frequency and density[7] of sleep spindles that are predominant in this stage are reduced and nREM2 bouts are highly fragmented.[8]

Sleep benefits memory in healthy young adults. This benefit has been demonstrated by comparing performance changes over an interval with sleep (e.g., 8p-8a) to changes over wake (e.g., 8a-8p).[9,10] Morning/evening control groups rule out the possibility that performance may be superior at one of these times of day.[10,11] Off-line changes in learning over sleep reflect *memory consolidation*, a process by which memory storage and retrieval becomes stronger and more efficient. Although memory consolidation may occur while awake, sleep-dependent memory consolidation (SDC) refers to the greatest portion of consolidation that occurs preferentially during sleep,[12] preventing consolidation from interfering with encoding of new memories.[9] Thus, sleep-based interventions may reduce deficits in memory in healthy aging (***Fig 1***, blue) and Alzheimer's disease wherein both memory and sleep are reduced (***Fig 1***, red).

We have examined SDC on a range of procedural and declarative learning tasks in healthy older and young adults. In short, our work,[13–15] in conjunction with others,[16] suggests that SDC for declarative memories (word-pair learning, visuo-spatial learning) is preserved, although reduced at times compared to young adults. However, SDC for motor procedural memories (motor sequencing, visuo-motor adaptation) is absent for older adults (for a review, see [17]).[13,18–20]

Innovation: Innovation is critical to the scoring of many NIH grants. For this reason, it is standard for an Innovations section to follow the Significance section. However, Innovation is not required of F31/32 proposals. As such, **an Innovations section is not required.**

Nonetheless, there are reasons to consider including it anyway. Those reasons are:

1. Although not required, many reviewers are so used to expecting NIH work to be innovative that they may be biased towards proposals with clear innovation.

2. If two equally good proposals are side-by-side and one has clear innovation, that one will be scored higher.

Given this, I recommend including a short innovation section, pointing to 2-3 ways in which your research is innovative.

It may seem easy to write out a statement of why you think your work is innovative. If that is you, then you are likely going to undersell your innovation. Or, you may think "My work isn't innovative.... I just do a follow-up project to the work of my advisor." You too are likely underselling the innovation of your work. The best way to find your innovation is to carefully read the NIH definition of Innovation. The criteria is really defined with a series of questions:

> The NIH criteria for Innovation: "Does the application challenge and seek to shift the current research or clinical practice paradigms by utilizing novel theoretical concepts, approaches or methodologies, instrumentation, or interventions? Are the concepts, approaches or methodologies, instrumentation, or interventions novel to one field of research or novel in a broad sense? Is a refinement, improvement, or new application of theoretical concepts, approaches or methodologies, instrumentation, or interventions proposed?"

So, consider your proposed work. Does one (or preferably more) of the following describe your work?
- The proposed work seeks to shift the clinical practice paradigm.
- The proposed work seeks to shift the current research paradigm.
- The proposed work utilizes an established construct from the field X and applies it as a novel application to field Y.
- The proposed work applies a novel technique to the field of X.
- The proposed work uses a novel approach to the question of X.
- The proposed work provides a novel application of {method} to the field of X.

Hopefully you now have chosen 2 or more of these bullets that you can construe as related to your project. Write those bullet points on your paper. Bold the text.

Now after each statement, write a few sentences of support for them. This should give you a strong Innovation section!

Here is an example of what that might look like:

* * *

2. Innovation

This proposal seeks to introduce a novel concept to the field of cognitive aging. The concept of SDC has gained wide support from behavioral and neuroimaging studies of young adults[48] and electrophysiology studies in animals.[49] Yet, the interaction between age-related changes in sleep and memory has received little notice. **Our work**[13] **was the first to assess SDC in older adults**, spurring pockets of studies in support of those findings.[18,50] Still lacking is an overarching view of the cognitive function of sleep in aging, primarily due to insufficient data.

This proposal seeks to refine approaches for studies of sleep and cognition. This field has taken a myriad of approaches (training to criterion,[13,29] statistically removing[18]) to ignore the role of encoding capacity in assessing SDC. Our results would point to the need to better control encoding strategies and measure encoding outcomes. We also provide a novel theory of SDC, encompassing multiple forms of learning, that will be important for the field.

This proposal seeks to shift clinical practice paradigms. We will define conditions for which sleep may account for age-related deficits in performance. Subsequently, sleep may be considered a target for treating deficits in these domains in individuals with Alzheimer's disease and other diseases where sleep is also impaired.[2,3] As an example, interventions to enhance declarative learning for individuals with Alzheimer's disease could be paired with sleep-based interventions to enhance the efficacy of the intervention, particularly early in the disease.

This proposal seeks to introduce a novel technique to the field of cognitive aging. High-density polysomnography (hd-PSG) is beneficial for understanding changes in the topographic distribution of slow waves and spindles[55] and the origin and propagation of these and other sleep features.[56] Only two studies have used hd-PSG in healthy older adults – one of which did not address age-related changes[57] and the other had no measures of behavior.[58] **Thus, the proposed studies will provide novel insight into age-related changes in sleep microstructure and the functional consequences of these changes.** Additionally, our pilot study and

Approach: This is clearly the 'meat' of the proposal. It's where you will say exactly what you plan to do to address your aims and how. For R21s and R01s, the score on this section is weighed heavier than any other section so it is a strong predictor of the overall score. It likewise drives the score for an F31 although perhaps the correlation (between approach score and overall score) is a little less steep. As much as an F31 is about the candidate, at the end of the day, reviewers are used to evaluating science and having good science is Indicative of the candidate's quality, so this is likely where you will spend a lot of time and get a lot of feedback.

Although you will have about 3.5 pages for the Approach, there is a lot that you will need to cover – methods, analysis, and figures of procedures, data, or predictions. So it is important to use your space, and words, wisely. What follows is a framework which may help you build the Approach.

1. First, type this skeleton in your document:

APPROACH
    <u>Overview of the Approach</u>
    <u>General Procedures</u>
    <u>Aim 1 [paste your aim]</u>
        Rationale
        Method
        Predictions and Analysis

Aim 2 [paste your aim]
    Rationale
    Method
    Predictions and Analysis
Aim 3 [paste your aim] (optional)
    Rationale
    Method
    Predictions and Analysis

It is possible that another skeleton may fit you work better. For instance, you may essentially have a single procedure that will answer three questions (aims). So your framework may look like:

APPROACH
  Overview of the Approach
  Method
    Participants
    Justification of sample size
    Tasks
    Justification of tasks
    Procedures
  Aim 1 [paste your aim]
    Predictions and Analysis
  Aim 2 [paste your aim]
    Predictions and Analysis
  Aim 3 [paste your aim] (optional)
    Predictions and Analysis

2. Now, let's fill it in. (Note: We will use the first framework in this example, but adjust to the framework that best fits your work.)

Overview of the Approach. This is where you will provide some basic information to justify or explain your strategy. I find it particularly helpful to include a diagram which clarifies the relationship between the aims. Or, perhaps you chose to include some "if… then…" statements. In other words, you are explaining your approach by saying "IF this (something that is predicted by the literature) is true THEN this (something you plan to test) will happen." This section does not have to be long but it can be helpful. You may come back and add stuff to this section after you've written the rest of the Research Strategy as this is a section I use to address potential questions. Illustrations are not required and may not make sense for all proposals, but a few examples follow.

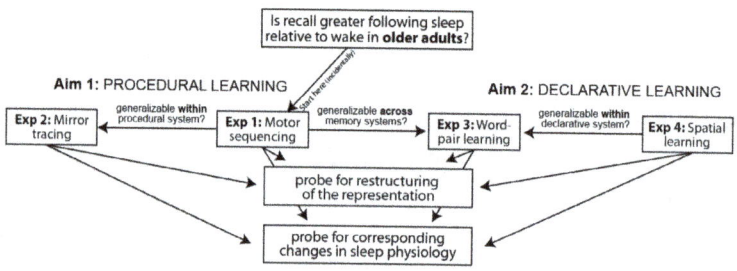

Figure X. This is an example of a figure used in the Overview of the Approach explaining the relationship between the overarching question (at the top) and the two aims and how these are addressed with parallel thinking (the bottom boxes).

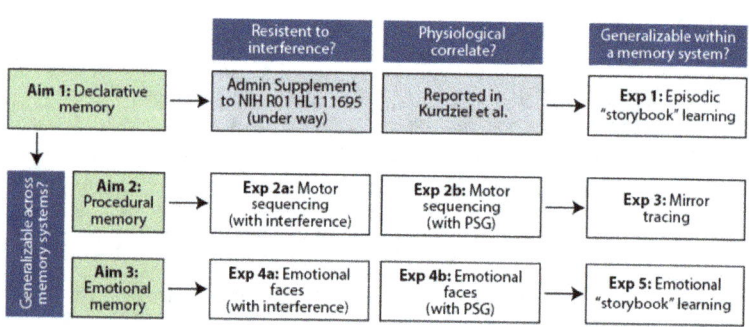

Figure X. Another example of how a figure can be used to illustrate the relationship between Aims (and experiments).

* * *

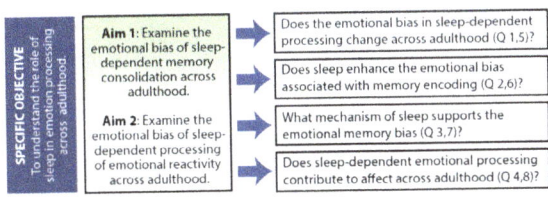

Figure X. Another example of how a figure can be used to illustrate the relationship between the overarching aims and the specific aims and the parallel experiments run under each aim.

General Procedures. This too is not required but for some procedures (see, for example, the second framework illustrated above which has a single methods section in place of this), it may make sense to give an overview first. But if you have a redundant structure to your studies (even if tasks or interventions or models or whatever will vary from aim to aim), then a General Procedures section may be for you.

Break this section into subsections as needed. Depending on your field, these may be:
- Participants or animals or whatever media
    o Sample size justification
- Questionnaires/tasks/interventions
- Protocol or procedures (i.e., the timing at which things will take place)

Aims (for each Aim)

In general, I recommend breaking the Aims section, into the following subsections. Again, this structure may not work for all proposals but you should consider whether some other sub-heading format might work for you.

Rationale: You may notice that you give some justification for your specific aims in the Significance, but you do not need to go into great detail there. Instead, you can justify this particular aim and your approach to this aim here. This is generally about 1 paragraph. You might draw on those 1-2 sentences you used in your Specific Aims statement to get you started.

Method: This is where you will describe your specific experimental method. Some studies have multiple experiments under a single aim, so you might need to separate these out. Of course you will not have space to go into great detail with methods. That is true for everyone and reviewers will understand this (as long as you fill all 6 pages – if you leave white space, they may be at liberty to ask why you did not give more detail!). But you should give enough detail that the reviewers are convinced you thoroughly thought through the project.

Predictions and Analysis: You have, of course, already formulated your predictions: these are the hypotheses you stated in the Specific Aims page. Here, you can start by restating that prediction. Then, you move to explaining how you will analyze the data to evaluate your hypothesis. This is really critical: what variables are you extracting from the data to address your hypothesis, and what is the statistical approach to determine the answer?

Potential Pitfalls and Alternative Approaches: Although brief, this section is a place where you can consider potential concerns of reviewers. Here you recognize what those potential concerns are and provide an explanation of why you chose the current path nonetheless and provide your alternative strategy should your plan 'not work'.

Start by making a list of potential things you think reviewers will scrutinize. This may be easier after you have the proposal fully written and get (honest) feedback from your sponsor, friends, or mentors. Be as curmudgeonly as possible, and consider big and small issues that are apt to come up - will they believe it is feasible to answer your question with the approach you propose? How might it go wrong? How might a different lab with different tools have solved the problem and why might your method be superior? Are there some critiques that you say - well, yes, that is a problem but this is nonetheless a good study?

With this list of questions, choose three or so that are most likely to come up and most legitimately concerning (again, ignoring whether they are big or small issues). Write each of these out. Now, after each, provide an explanation - Why is it not a concern? Why is your approach superior? What will you do if your planned approach does not work? Provide a concise response for each and tie these into a tight paragraph (or possibly a short paragraph for each concern if you have space).

* * *

Feasibility: Given that reviewers are interested in whether the work is feasible within the timeframe you propose and whether the work is reasonable given the stage of your career and research experience you have, it is worth emphasizing feasibility in a brief section.

Points of feasibility will vary based on the research area. However, here are some general points you might consider in this section:

- The procedures are feasible: Point to similar work that the candidate and/or sponsor has done which suggest that, likewise, the proposed work is feasible. In some cases, you may have some pilot data and, even if the quantity is too limited to support the effects, it may support the feasibility.
- Sufficient resources are available: Are the specific resources that you have access to that will facilitate the efficiency of data collection or other aspects of feasibility? This is important to emphasize here (and in the Facilities and Resources section).
- You can achieve your research goals in the appropriate timeline (which you can transition with to the next section).

Timeline: This section is largely addressed with an illustration. The text simply needs to point to that. For example:

*The distribution of the proposed work across the funding period is illustrated in **Fig X**.*

Success will be measured by meeting the goals laid out on the timeline. More

importantly, peer reviewed papers will mark the completion and impact of the work.

Then insert the figure developed in Chapter 3 (Testing Your Scope: Sketching out your timeline).

CHAPTER EIGHT

Training in Responsible Conduct in Research

Training in research ethics and the responsible conduct of research is important and NIH recognizes this. Responsible Conduct in Research (also called RCR) training can come about both formally (on-line or in-person courses) or informally (learning from mentors). You should include a description of both types of mentoring.

The up-side to this section is the NIH makes it very clear what they are looking for with regards to Responsible Conduct in Research. See here: https://grants.nih.gov/grants/guide/notice-files/NOT-OD-22-055.html This chapter will instruct you as to how to achieve these requirements.

Format requirements:
1 page limit
Standard format (.5 inch margins, 11-point Arial font)

---

1. As always, start by formatting your page as described above and insert you section title:

    TRAINING IN THE RESPONSIBLE CONDUCT IN RESEARCH

    - Now give yourself two section headers. I recommend two sub-headers under these if they are relevant:

Past Training in the Responsible Conduct of Research

Planned Training in the Responsible Conduct of Research

2. Now fill in these sections.

Under the "Past Training" section, develop a paragraph describing your experiences with RCR training. While it may be tempting to try to integrate courses you've taken which may have discussed ancillary topics, focus on formal training first (have you taken and ethics class? CITI training modules?). You can also include formal training that is relevant. Ideally, you have had ethics training at each career stage (undergrad, post-bach research position, etc).

Now here is an example of this section (note that this is from an F32, a post-doc application) but the requirements are the same.

Under the "Planned Training" section, you should identify and describe the formal training opportunity that you will take advantage of to meet this requirement. Most campuses offer a course or workshops which are designed to explicitly meet these requirements. While previously this course had to be in-person, these rules have recently changed and some on-line components, if they are discussion based, are allowed. However, importantly, you still cannot rely solely on on-line training. As stated by NIH, *"video conferencing should not be the sole means for meeting the requirement for RCR instruction, and a plan that employs only video conferencing will not be considered acceptable…"*

After you have identified the opportunity, describe the following:

Format of instruction: The requirement is that this training be discussion-based whether in-person or on-line (and how much time is spent in-person and on-line will be important to clarify).

Subject matter: Describe what the course will cover. This may be the weekly topics from the syllabus. Topics of interest to NIH are:

- conflict of interest – personal, professional, and financial – and

- conflict of commitment, in allocating time, effort, or other research resources
- policies regarding human subjects, live vertebrate animal subjects in research, and safe laboratory practices
- mentor/mentee responsibilities and relationships
- safe research environments (e.g., those that promote inclusion and are free of sexual, racial, ethnic, disability and other forms of discriminatory harassment)
- collaborative research, including collaborations with industry and investigators and institutions in other countries
- peer review, including the responsibility for maintaining confidentiality and security in peer review
- data acquisition and analysis; laboratory tools (e.g., tools for analyzing data and creating or working with digital images); record keeping practices, including methods such as electronic laboratory notebooks
- secure and ethical data use; data confidentiality, management, sharing, and ownership
- research misconduct and policies for handling misconduct
- responsible authorship and publication
- the scientist as a responsible member of society, contemporary ethical issues in biomedical research, and the environmental and societal impacts of scientific research

Frequency and timing: You must get RCR training once in each career stage (undergrad, post-bach, grad school, post-doc) and it must be taken at least once every four years. Regardless of when you have it last, you must take a course during the fellowship period. Most often, you will find access to a course offered for a semester and it will take place weekly. Some campuses, however, do monthly workshops and you can accrue hours by attending these.

Faculty participation: The course should be taught by faculty. Make sure this is clear

Duration of instruction: This is where you describe how often the course/workshop will meet and how long those are. It is important that the duration of instruction in total be at least 8 hrs but is typically more (10-15 hrs)

EXAMPLE: In this example, note in particular how the formatting (underling) draws out what the reviewers are instructed to look for.

TRAINING IN THE RESPONSIBLE CONDUCT IN RESEARCH

<u>Past Training in the Responsible Conduct of Research:</u>
As an undergraduate student at State University, I completed Introduction to Ethics. This was a 1-credit in-person discussion-based course, offered once weekly for approximately 1 hour each week for 13-weeks and taught by Dr. Christopher Dread. Training focused on topics of value, morality, and theories of right/wrong behavior. The course was designed to help students develop their own ethical positions using seminal theoretical frameworks by reviewing and critically analyzing modern ethical issues, including those of human research misconduct, animal maltreatment, and topics of diversity and representation.

As an undergraduate, I also completed the online CITI certificate in Biomedical Research which included topics such as: Belmont report principles, history and ethics of human subjects research, IRB regulations and review process, informed consent, and conflicts of interest. As a <u>post-baccalaureate</u> trainee, I completed online certificates in Social and Behavioral Responsible Conduct of Research, Human Research, and Good Clinical Practice which included the following topics such as: authorship, conflicts of interest, research misconduct, and reporting adverse events. As a <u>graduate student</u>, I completed the online certificate for training of Human Research: Group 2 which included the following topics: populations requiring additional considerations and/or protections, history and ethical principles, assessing risk, and informed consent.

<u>Planned Training in the Responsible Conduct of Research:</u>
I will enroll in the course Research Ethics (BIO 891) in the first year of the award. This course is designed to meet the objectives of NIH RCR requirements.

<u>Format of instruction</u>: This is an **in-person** course facilitated by readings and **discussions** of the core topics.

<u>Faculty participation</u>: This course is taught by Dr. Doolittle, a professor in the department of biology and an NIH-funded researcher.

<u>Duration of instruction</u>: This course takes place once/week for 75 mins for 13 weeks (~16.25 hours).

<u>Subject matter</u>: Specific topics covered by this course include **Research Integrity**: When to report misconduct and how to know misconduct when you see it; **University Resources and Policies:** Information about the ORI and how to access the proper resources to report misconduct; **Mentoring:** Understanding the proper expectations and interactions of mentorship; **Proper Data Collection & Management:** Lessons for the preservation of privacy and ethical experimentation; **Authorship and Review:** Guidelines and rules for deciding authorship and gauging the importance contributions to research projects and ethical considerations of the peer review process; **Humans in Research:** Understanding the history and instatement of important ethical principles outlined in documents such as the Belmont report; **Animals in Research**: Exploring the IACUC and ethical principles concerning the use and treatment of animals; **Competing Interests and Collaborations:** What constitutes financial, personal, and other competing interests and considerations of ethical collaborative relationships.

<u>Frequency and timing</u>: This course will be taken in the first year of the award. In subsequent years, additional RCR training will include: (1) CITI refresher courses; (2) informal training via one-on-one with mentors; (3) hands-on experience in developing and maintaining IRB protocols.

CHAPTER NINE
Biosketch

The NIH Biosketch provides an opportunity for the reviewer to know more about you as a researcher. For fellowships, this is a critical section as the awards are just as much, if not more, about the *person* than the *research*. So, whatever you do, do not take this section for granted.

Formatting your Biosketch

NIH has a Biosketch template that you will use and the template is often paired with an exemplar Biosketch. This formatting changes every few years, so when you begin, you may want to check for any changes. For now (5/2022), these are the current documents are here:
https://grants.nih.gov/grants/forms/biosketch.htm

BE ABSOLUTELY SURE that you use the fellowship format and sample NOT the non-fellowship options.

BE ABSOLUTELY SURE that your biosketch is 5 pages or less.

To be straightforward, the Biosketch is one of the most important documents as it characterizes your successes. But, these may feel out of your control. In particular, **reviewers want to see publications**. In general, more publications will predict a higher your score of the candidate. If you can do anything to push publications along before you submit this grant, it may be worth your time.

The top of the template will ask you some seemingly straightforward questions but they often end up more confusing than you may think. Here I attempt to preempt some of your questions.

- **Name:** Obvious. But be sure to enter in the order of: Lastname, Firstname Middlename
- **eRA Commons user name:** As a graduate student, it is very likely that you do not have an eRA Commons username yet unless you are funded by your mentor's NIH award or have previous received or applied for NIH fundings. If you do not have one, contact your campus grants office who can generate this unique identifier for you.
- **Position title:** Your title is likely "Graduate Research Assistant"

Next there is a table for you to list your professional education and training experiences. As a graduate student, this may just be your Bachelor's institution but you may already have a Masters degree which would, of course, also be listed. If you completed a relevant certificate program or other relevant training **program** (with an emphasis on program, meaning it was a formal training experience), list that here.

- **Institution and location:** Name of the place that you were trained. It is interesting that many of the Samples out there do not list 'location' as instructed. With a name like "University of Chicago", it is not necessary to state the exact location. However, if you go to Lesserknown College (a small college or college without a reputation or name that specifies its location), it is worth clarifying its location here (you will have plenty of space).
- **Degree:** Of course BS/BA or MS/MA are obvious responses. If you took a short course at a national lab, you could state "certificate in X". It is also acceptable to leave this blank if it was a non-degree training experience.
- **Start date/End date:** Hopefully these are obvious. Overall, reviewers are interested in your timeline. (If you have gaps in this timeline that are worth explaining, this would go in your personal statement.)
- **Field of study:** This is asking what you have a degree in. In other words, this should not reflect the niche that you researched or focused on, it is as broad or narrow as your degree field.

* * *

Sample Education/Training table

INSTITUTION AND LOCATION	DEGREE (if applicable)	START DATE MM/YYYY	END DATE MM/YYYY	FIELD OF STUDY
Workmore College Hometown, PA	BS	08/2012	05/2016	Biology
University of Bliss Anytown, MA	PHD	08/2016	(in progress)	Neuroscience

The **personal statement** explains why YOU are the right person for THIS project. YOU: This statement should be clear what you bring to this project, including a motivated enthusiasm for this work. THIS project: Your personal statement should be adjusted to the purpose for which you are using it. You will tailor the personal statement to the project that you are including it with, in this case, the research you propose in the Research Strategy section and your general goals for the fellowship. Make specific connections between your interest and experience and what you plan to do.

This is the section to explain many things about you that do not fit elsewhere. It is also **written in the first person** and it is completely expected that you are marketing yourself as the right person here. In other words, braggery is not only appropriate, it is expected!

If you are an innately humble person, pretend that you are writing about someone else with your accomplishments. Word it as if you are writing their nomination for a Nobel prize. That, countered by your humbleness, will hopefully bring you to the right level of confidence in pointing to your strengths. However, do not go to the point of arrogance. Most of your bragging should be around clear statement of facts without using subjective terms (e.g., "I am particularly good at …").

Although there is no strict format for this section, here are some pointers:
  Length: ½-¾ page (about 300-400 words)
  Format:
- Start with a strong statement of who you are
- Then state your long-term objective (career goals, intended career trajectory)

- Follow that with a few objective statements of your experiences. If you have specific research experiences, this is a place where you can describe them (succinctly of course). But save your scientific accomplishments for section C.
- Then move to describing how you have come to this objective. I.e., what experiences have you had that make you well-suited and/or certain of this pursuit.
- Then build on how this will bring you to achieve your long-term objective.

Points to address: Sprinkled throughout your statement, be sure to address…
- If there were **gaps in your timeline** or an evident 'jump' in your career trajectory, be sure this is explained. This can include family/parental leave.
- Although you will find that section D gives you an opportunity to present your scholastic performance, it is fair to point out particular strengths that your reviewer may glance over. For instance, perhaps you took an ample number of classes in statistics or programming. If these skills are important for the work you propose, certainly be clear that "As part of my undergraduate coursework, I took 4 classes in advanced statistics which will be an asset to my proposed project."
- But, you will also note that section D asks you to give coursework and grades. The Personal Statement is your opportunity to confront any gaps. For instance, if you did not start out strong as an undergraduate, you may state the reason here, e.g., "Although I learned the hard way that the Plant Biology major was not the best fit for me, this challenge is what brought me to the research area that I have a passion for and my most recent academic performance is truly reflective of my abilities."
- It may be good to include a **COVID-impact statement**. This is a brief paragraph that explains how your productivity (research progress, publications) were slowed by the pandemic.

- DO NOT use figures, tables, or graphics. I also recommend <u>not</u> using bullet points here. This should be a narrative text.

You can start by making three sub-headers in this section and list your accomplishments in <u>chronological order</u> in each category.

- **Positions and employment:** List any relevant previous positions. For grad fellowship applications, this may be a research assistantships, teaching assistantships, or internships, or paid employment in relevant settings. Be sure to list the relevant start and end dates (by year is typical).

> **Box 3. On being notable – and remembering**
>
> Often times, students are more successful than they remember being. You may present your research at a small on-campus conference and forget about it when you use your CV for an application a year later.
>
> To prevent this, open your CV often. Every time you have an accomplishment – a poster presentation, a talk, your join a committee, etc. – add it to your CV right away. Don't wait until you're asked for your CV to update your CV.
>
> I refer here to your CV. This will be the document most used for job applications and many award applications. You should then use that to fill in your biosketch and to update your biosketch when needed.

- **Other experiences and professional memberships**: If you are a member of your research society or other professional membership, list those here as well as the dates of your membership (e.g., 2010-2012, 2012-present)

- **Honors:** List any scholarships or awards you received.

> **Sample Positions and Honors section**
>
> **B. Positions and Honors**
> **Positions and Employment**
> 2014-2016   Undergraduate Research Assistant, Science Laboratory, Workmore College
>   PI: Dr. Still Morework
> 2016-   Graduate Research Assistant, SciFi Laboratory, University of Bliss
>   PI: Dr. Domore Work
>
> **Other Experience and Professional Memberships**
> 2015-2016   Member, Psi Chi
> 2016-   Member, SciFi Research Society
>
> **Honors**
> 2012-2016   Town Hall Scholarship, Hometown Members Association
> 2016   Best Thesis Award, Department of Biology, Workmore College
> 2017   Graduate Student Travel Award, Neuroscience Program, University of Bliss

The format of this section will vary depending on your experience. Described here is the format expected for a pre-doctoral fellowship application but pre-doc even have varied experience.

This section can have up to 5 subsections. Typically, experience at this point in your career is best segmented by era of your training: high school research (if any), undergraduate research (if any), post-bach research (if any), and graduate research. Under each, you can list any products (publications or conference presentations). The example given by NIH is a very good exemplar (see below).

A few notes as you pull together this section. First, you likely do not have high school research experience and do not be intimidated by this, it is the exception more than the rule. Certainly don't stretch your experiences to come up with something to put there. Second, if you had two rather significant and distinct undergraduate lines of research, then you may choose to categorize these by research areas rather than by career timing.

IMPORTANT: Note that "in press" and "under review" papers and conference presentations are accepted for graduate student trainees. These may not be worth as many "points" but it is nonetheless a good idea to include these particularly if you are weak in products. In general, given that you have a limited number you can report, report publications over in-press or under review work. But your advisor may not be aware of this so they may question it. Refer them to the NIH forms E webpage for clarification.

In this section, you will provide the reviewer with a list of classes you took during your training as well as your grade. These should appear in chronological order and include <u>all</u> graduate and undergraduate classes. You should also provide an explanation of the grading scale: 0-100, A-F, or 0-4.0. This is typically a description provided on your transcripts if you are uncertain.

Note that you do not need, nor should you include, the course number (i.e., it does not help the reviewer to know it was BIO 210 or PSY 582).

This section should appear as a table as illustrated in the NIH Biosketch Fellowship Sample:

SCHOLASTIC PERFORMANCE

YEAR	COURSE TITLE	GRADE
	Statesburg College	
2011	Molecular Biology	A
2011	Chemistry Principles	A
2011	Population Biology	A
2011	French Dialogue	A
2011	Physics I	A
2011	Learning and Thinking	B
2011	Linear Statistics	A
2011	Animal Behavior	A
2012	American Literature	A
2012	Physics II	A
2012	Organic Chemistry I	A
2012	Organic Chemistry Lab I	A
2012	Developmental Neuroscience	A
2012	Pathology	A
2012	World Anthropology	A
2013	Organic Chemistry II	B
2013	Organic Chemistry Lab II	A
2013	Human Genetics	A
2013	Biology & Culture	A
2013	Behavioral Neuroscience	A
	University of Paradise	
2014	Molecular Neuroscience	A
2014	Seminar in Neuroscience	A
2014	Advanced Statistics	A
2015	Systems Neuroscience	A
2015	Seminar in Neuroscience	A
2015	Neuroimaging	A
2015	Ethics in Biological Research	A

All courses are graded on a A-F range with passing reflected as a C+ or better.

Given that this list of coursework can take up quite a bit of space, I will end by reminding you that **the biosketch must be 5 pages or less.**

CHAPTER TEN

Applicant's Background and Goals for the Fellowship

The Background and Goals section is one of the most important sections in your application. This section tells reviewers what your background is in a narrative form, what you want to do with your career, and how you will use this fellowship to get there.

Incredibly important in framing this piece is this: **You must be proposing a course of training that is different than if you were not on a fellowship.** If you will do the same research, go to the same conferences, and take the same classes whether or not you get this fellowship, your score will be low.

You also must have a clear goal as to what you want to do (e.g, "I want to be a professor at a Research 1 University, with research focused on sleep and psychopathology in older adults.") **and what you need to do to get there and how you will do that.** For instance: "To do this successfully, I will enhance my existing training in clinical neuropsychology and gain hands-on experience in sleep research techniques.."

The F31 fellowship is really intended to fund the person as opposed to the research: Funds cover your stipend, not the research expenses. You provide a research plan to show them that you can formulate a strong research plan and use the stipend wisely, but they really are concerned with your trajectory – do you have a plan for where you are going beyond your degree.

Take careful note, before you start writing, of the page limit. Six pages is pretty substantial and breaks down to about 2 pages per recommended subsection below. This tells you two things. First, being as long as your Research Strategy, reviewers will care just as much about this as your Research Strategy. So, **develop this section with equal attention and care**. Second, just like the Research Strategy, use your space to the fullest. Before you start writing, think about how you will use the space to its fullest. But, do not be overwhelmed, take it one section at a time!

> **Format requirements:**
> 6 page limit
> Standard format (.5 inch margins, 11-point Arial font)

1. As always, start by formatting your page as described above and insert you section title:
   CANDIDATE'S BACKGROUND AND GOALS

Now give yourself three section headers:
      A.   **Doctoral Dissertation and Research Experience**

      B.   **Training Goals and Objectives**

      C.   **Activities Planned under this Award**

2. Now you can develop these sections.
   A. Doctoral Dissertation and Research Experience (~1.5-2 pages)
   In this section you will list and briefly describe your prior research experiences (note, not coursework). Having done your biosketch, you can take the accomplishments listed there and now put them into context here. You should also include your doctoral research. You may break this into subsections of:
   - Undergraduate Training
   - Post-baccalaureate Training
   - Graduate Training

Or, perhaps you had multiple experiences as an undergraduate, in which case you might format by project. **In either case, bullets individuating the experiences is helpful.**

* * *

B. Training Goals and Objectives (~1.5-2 pages)

This section should provide a clear vision (before you panic, see Box 4). What is your career goal, and how will this award help you reach that goal? Now that you have mastered the Research Plan, think of it like this: your training goal is like your long-term plan and your objectives are like your specific aims. So, before you start writing this section, complete the following:

My long-term career goal is:
_____

Specific objectives of this award that I will take to reach this goal are:
1) _____
2) _____
3) _____
(of course you could have more than three…)

Now you can get started. I recommend breaking this into subsections with clear labels in bold.
- Career goal
- Fellowship training goals

Now develop these sections.

> **Box 4. Don't be afraid of the plans you are making!**
>
> Don't be afraid of defining what you will do *for the rest of your life* when writing the Background and Goals section. This section, and the whole proposal, are intended to demonstrate you have a plan for success and a past that makes it likely you will succeed on that path. **Consider this your working hypothesis as to what you want to do going forward.** Importantly, you will not be held to the plan exactly as stated. But this should be a reasonable guess as to what you current plan to do with your career. It is an excellent exercise to think about what that may look like.

Career objective: In this section, start by addressing the primary objective. For instance, "My long-term career objective is to be an independent investigator in a university setting." But there is more than this – build on what type of research setting you envision. What is the research direction you would follow in your lab? You do need to reconcile this with your research plan. One option is that your long-term research directly follows on the type of research you are pursuing for your proposal and dissertation. Particularly if this is novel work that is relatively independent of you advisor's work, this is likely the way you will want to go. A second option is that your research is giving you foundations that you will integrate with something else to establish your own line. This is a recommended option if your work is close to your advisor's work – you should recognize this and be clear how you will establish your own unique research scope. **A strong clear statement should start this section such as: "My career objective is to be an independent research investigator at a research university with a research focus on sleep in childhood psychopathology."**

Next, recognize your full trajectory needed to get there. In most cases, you will need a post-doc prior to an independent career. Recognize this. Discuss what gaps you will fill with a post-doc following your doctoral career.

Next, you should establish your career objective beyond research. This might be what we think of as "Broader Impacts" for an NSF proposal. How will you contribute to your field through mentoring and/or outreach? It is best to build on something you have already started. If you are currently doing mentoring of undergraduate students in your lab, integrating similar mentoring into your plan is a good way to go.

C. Activities Planned under this Award (~2-3 pages)
This section should have two subsections:
1. Description of Activities
2. Timeline for Activities

In this section, provide a detailed plan as to what you will do during this award aside from your research. You might bullet this section by the particular technique, theory, and/or conceptual approach you want to acquire (and you should have at least 3 of these that you could list).

Consider in particular **research fundamentals** that reviewers may now see as critical for up and coming researchers – e.g., statistics, computer programming, computational modeling. Also consider **basic career skills**, such as manuscript preparation, speaking skills, and grant writing. Of course you should have particular **research techniques** you seek to acquire as well.

* * *

For each of these, state the gap and what you will do to fill the gap. For some, you may simply just take advantage of opportunities. For instance, to practice speaking skills, you will take advantage of opportunities to speak at lab meetings, departmental colloquium, and conferences. But for other key techniques or skills you should have a more significant plan – consider whether you could attend a short training course somewhere or take a class offered on campus. Provide this for each of your gaps.

Now develop a timeline for these activities. Given that you also need to indicate how much time is spent on each activity, these might be nicely summarized in the table as well. A clear yearly plan is best illustrated with a table or figure as follows:

3b. Timeline and Activity Allocation

Collectively, the benchmarks of this training plan and time to be allocated to each activity are in Table 1.

Table 1. Timeline and time allocation of proposal training activities.

Activity	Mentor(s)	Year 1				%	Year 2				%
		Su	F	W	Sp		Su	F	W	Sp	
Individual Meetings with Mentors (Y1 = 9%; Y2 = 11%)											
Sponsor meetings (weekly)	RB	X	X	X	X	6	X	X	X	X	6
Co-sponsor meetings (bimonthly meetings)	TT	X	X	X	X	3	X	X	X	X	5
Research Group Activities (Y1 = 3%; Y2 = 3%)											
Research team meetings (weekly)	RB	X	X	X	X	3	X	X	X	X	3
Sleep Science Training (Y1 = 7%; Y2 = 4%)											
Sleep science lab training (biannually)	RB		X		X	2		X		X	2
Sleep science seminars/trainings (monthly)	RB	X	X	X	X	2	X	X	X	X	2
PSG scoring	RB	X	X	X	X	3					0
Neuroscience Training (Y1 = 26%; Y2 = 15%)											
Psych 330 online course	N/A			X		6					0
Psych 731 in person course	N/A				X	10					0
MRI training	TT	X	X	X	X	10	X	X	X	X	10
MRI observations	RB					0	X	X	X	X	5
Developmental Science Training (Y1 = 8%; Y2 = 8%)											
Cognitive task training	RB	X	X	X	X	3	X	X	X	X	3
Developmental Science seminars	RB		X		X	2		X		X	2
Independent readings	TT	X	X	X	X	3	X	X	X	X	3
Statistical Analysis Training (Y1 = 12%; Y2 = 4%)											
Statistics consultations	RB	X		X		2	X				2
"Analyzing Intensive Longitudinal Data" Course	N/A	X				5					0
"Multilevel/Mixed Models Using Stata" Course	N/A			X		3					0
Conference statistical workshops	RB	X				2		X			2
Grantsmanship and Writing (Y1 = 26%; Y2 = 43%)											
Grant writing workshops	RB		X		X	3	X		X		3
SRS Small Research grant development and submission	RB		X			5					0
Develop and submit NIH K01	RB				X	3	X	X	X	X	10
Prepare and submit manuscript 1	RB & TT			X	X	10					0
Prepare and submit manuscript 2	RB & TT					0			X	X	10
Responsible Conduct of Research (Y1 = 2%; Y2 = 5%)											

In-person RCR course	N/A					0			X	3
Additional online RCR courses (biannually)	N/A	X		X		2		X	X	2
Professional Affiliations and Engagement (Y1 = 4%; Y2 = 4%)										
Conference attendance and presentations	RB & TT	X	X			2	X	X		2
Professional and special interest group meetings	N/A	X	X	X		2	X		X	2
Mentorship (Y1 = 3%; Y2 = 3%)										
Mentorship of junior research assistants	N/A	X		X		3		X	X	3

Key: CRF = Center for Research on Families; ISSR = Institute of Social Sciences Research; PSG = polysomnography; Y1 = year 1; Y2 = year 2

Box 5: Using tables to highlight accomplishments and goals.

Consider using tables throughout the Background and Goals section to highlight key points. For instance, in section A (Doctoral Dissertation and Research Experience), a table may highlight: Research project, Techniques acquired, Coursework/Workshops, Outcomes (posters, papers). In section C (Activities Planned under this Award), a table could highlight: training gap/fellowship objective, coursework/workshops to meet objective, and mentor to support you in this objective. Remember to particularly highlight aspects that are above-and-beyond what you would normally do in your program.

Tables in this case do two things. First, tables can use smaller fonts! I recommend 9-10 pt font and not lower. Second, they draw out things that may otherwise get lose in the text. Third, it provides the reviewers a reminder when they go back to your proposal later (such as for discussion at the review panel).

Example table for section A:

Research project	Skills & techniques acquired	Other training	Outcomes
Undergraduate research			
Sleep in canines	Actigraphy recordings of sleep; Animal care protocols	R stats package workshop; Animal Behavior course; Sleep Grand Rounds	Smith et al. (June 2019). Let sleeping dogs lie: Hours slept by the average household canine. Poster presented at SLEEP annual meeting, Philadelphia, PA
Graduate research			
Circadian cycles of cats	REDCap; Questionnaires and survey data collection; Focus groups	Vet tech in-training certificate; Sleep Grand Rounds	Smith et al. (2021). Evil felines: Sleep/wake cycles in household cats modify cycles of their owners. *Acta Animalia*, 63:19-22.
Sleep of pet owners	Human sleep physiology; Statistical modeling	Sleep disorders course; Longitudinal Data Analysis course	Smith et al. (2022). My dog at my REM sleep: disrupted sleep in animal owners. *Canine Care*, 55, 11-15.

CHAPTER ELEVEN

Facilities and Other Resources

This section should convey to the reviewers that **not only are the facilities available** to do the work you propose **but that you are in the best place to do it**. This may seem like a tall order at first, but if you think more broadly and by category, this section can be developed successfully.

> **Format requirements:**
> no page limit
> Standard format (.5 inch margins, 11-point Arial font)

Since these sections describe the space and resources of your lab and department, it is certainly worth asking your sponsor for this section of their grant if they have a history of NIH funding. If they are not NIH-funded, ask your sponsor if there are NIH-funded faculty in your area that have similar needs that you might ask for this section for ideas as to what to include and descriptions.

Even with those in hand, think as carefully as possible about what is available to you as you may have different needs and be able to produce a more developed section. This section can be broken into subsections as follows. Not all of the following may be relevant to you, but browse the list, and be sure to read the description of each category carefully before deciding whether it is relevant.

Research Facilities

Laboratory space: Describe the size of your laboratory space (or spaces, as often labs are composed of multiple rooms) and how it is utilized (bench space, testing rooms, etc).

Office space: State the size of the office space, what it is furnished with, and where it is situated relative to the lab and/or department.

Core facilities: If you use campus core facilities of any sort (any facility housing equipment used in your research or related research), describe those in detail.

Animal care facilities: If you do research in animal models, provide a brief description of animal care facilities and support services (such as on-staff caretakers and veterinarians).

Clinical facilities: If your research uses clinical populations, chances are you rely on clinical facilities for screening, recruiting, or receiving patient populations. Describe these facilities – reception area, exam rooms, interview rooms, etc.

Research services:

- Shop services: Some departments or colleges provide shop services to build or repair equipment. If this is available (and you might look into whether it is if you are not already aware), describe these services: number of staff, equipment in the shop, examples of the work they do. Relatedly, some research, like chemistry, utilizes a professional glassblower to design, produce, and repair glassware. These services should be described as well.
- Technology services: You likely utilize campus technology services – they provide the capacity to network across buildings or rooms; they provide wi-fi access; free software including statistics packages, may be provided by your campus, and their staff may provide computer repairs when necessary. Describe these services in detail.
- Library system: You may be take it for granted, but you probably have an extensive library that you use for journal articles or books related to your research. Describe these. I suggest looking around for some formal descriptions of your libraries (in campus

brochures, websites, etc) for some concrete numbers to describe the space/size (e.g., # of books, sq ft, # of floors, etc).

Environment

Institute or centers: You and your lab may be part of an institute or center which provides support for you. For example, these affiliations may host faculty/student retreats, seminars, or training workshops. Describe these here. If you know of the source of financial support, provide that as well (e.g., "the NSF-funded Center for Lobster Research).

Department and college resources: You may receive support at the department or college level as well. Examples to consider: a graduate studies staff person who provides support for you, a seminar or colloquium series, or travel awards offered through the department or college.

Graduate school resources: At the graduate school level, there may be resources available to you as well. At this point in your career, perhaps you have not explored these but this is a good time to learn what is available that you might take advantage of (and build on these in the Background and Goals section). For example, your graduate school may offer travel awards, workshops on career development, or writing support groups. This makes your graduate school an asset that you could describe here.

Other Contributions to Success

Location: Perhaps the geographic location of your university is in itself an asset. For example, you may study a very specific human population and your large urban city is an asset to draw from. Or, perhaps you are near multiple other colleges and universities which provides additional support in terms of access to public talks or libraries. This may be worth pointing out.

Human subjects: If your research utilizes human research participants, you should describe resources available for recruiting those participants. For example, some places have a shared recruiting database or formal relationships with community partners.

Animal strains: If you do animal research, particularly with unique animal lines, consider whether your resources for this are unique and could be described as well.

CHAPTER TWELVE

Letters of Support, Recommendation, and Eligibility

There are two types of letters accepted with the NIH F31.

<u>Letter of reference</u>: As the name suggests, in the letter of reference, the writer provides a recommendation - supporting that you are a good student with a lot of potential for success. **Three letters of reference are required with the F31 application.**

<u>Letter of support (optional)</u>: If your research relies on the support of someone other than your sponsor or co-sponsor, you can provide a letter of support. Importantly, a letter of support is not required from a sponsor or co-sponsor. Rather, the letter of support would be from someone you would consider a collaborator. Or this may be someone who is perhaps providing access to their equipment, providing materials (e.g., a mouse strain), or serving as an area expert if your proposed research spans outside of the expertise of your sponsor and co-sponsor (be sure to critically evaluate whether they should serve as a co-sponsor). Many proposals do not have such commitments and, as such, **the letter of support is not required.** But there is also no 'cost' to adding one. So if you have anyone playing any substantive role besides your sponsor and co-sponsor, ere on the side of caution and include a letter.

LETTERS OF REFERENCE

Letters of recommendation are a critical component of the F31 application and can boost your application. However, letters can just as easily do nothing or, worse yet, hurt your application. Here is how those scenarios can play out.

The useless letters: Often, letters are useless. (In my personal opinion, this is 90% of letters.) Every advisor or mentor or former professor from a class you took is going to find something lovely to say about you and fill a requisite 1 to 1.25 pages. These aren't bad. They aren't going to hurt your score. But they also do nothing to bump you up in a reviewer's mind either as they are the status quo.

The harmful letters: Harmful letters are rare. And they are rarely blatantly bad letters. It is not common that a professor will agree to write a letter and then submit a poor one without giving you some advanced warning. If they do not have positive words to say, they would likely explicitly let you know (e.g., "I am happy to, but I will be honest that I do not have strong things to say about you based on your poor attendance in my class.") or implicitly let you know (e.g., "I am sorry, I will not have time, in the next 4 months, to write a letter.").

In most cases, the harm comes from red flags that they raise. For instance, a red flag may be a particularly short letter from a mentor who should know you particularly well (e.g., your undergraduate honors thesis advisor submits a paragraph that you did "fine"). Other red flags come from the content of the letter being either vague or focused on your personality as opposed to your science and skills. Small red flags in otherwise positive letters can be harmful given that such a large number of letters are positive (the useless letter is always glowing), you are now set apart by this red flag that distinguishes you from the rest even if minor.

One additional note, is that letters could collectively be harmful if all 3-5 letters were from writers that didn't seem to know you well and/or from less ideal letter writers (see "Who should letter writers be?" below). This raises the worry that you either did not get to know your professors or that they may have been, for reasons one can only wonder about, unwilling to provide letters.

The helpful letters: Helpful letters are those that stand-out from the useless letters by providing specific levels of detail that either give a better understanding of the candidate or provide unique information about you that help you stand out amongst the rest of the applicants. One area that letters can help with is by addressing gaps in your record. For instance, if you had low grades early in your undergraduate career, your undergraduate research advisor may be able to comment on how those problems you had in adjusting to college life were so clearly overcome and do not reflect the researcher they knew you to be. Or, a letter pointing to the fact that the research you have successfully accomplished in this first year of graduate school was particularly impressive given that you had traveled frequently to visit a sick family member. Collective comments across your letters regarding your particular investment in some outreach program, is another example of letters that could boost your application.

<u>Three to five letters of recommendation are allowed</u> and I strongly recommend you request five. This way if anyone fails to meet the deadline, you still meet the minimum.

Let's start with who the letter writers cannot be: **your sponsor and co-sponsor (i.e., your grad advisor and co-advisor) CAN NOT be letter writers**. The idea is that they are exhibiting ample support in other sections of the application.

Additionally, with few exceptions, I advise that the **letters <u>not</u> be from post-docs or non-academics**. The reason being, that post-docs are from labs that had PIs and you should ask the PI of that lab for the letter and the view from the post-doc should be somewhat redundant with that. Sometimes letters are co-written by a PI and the post-doc which is fine.

As academics tend to respect the experiences held inside academics a bit more than those in industry or otherwise, it may be best to stick with academic letter writers (whether this is an problem with academia is certainly worthy of discussion, and a topic for another book). Likewise, there is a bit of snobbery that may have reviewers looking at the credentials of the reviewers as a whole, and respecting those who are full professors more than assistant professors and not giving credence to lecturers. I have been on a grant panel where one candidate was described as "oh, the one that had only assistant professors as their letter writers?" Again, it may not be a measure that you or I agree with, but be aware that these biases exist.

The <u>exception</u> may be that if you did an internship or post-bac research experience in an industry setting, certainly include those internship supervisors as letter writers if they know you and your scientific research ethics/abilities.

So who should these letter writers be? Here is a list of <u>who to consider for letter writers</u>:

- **Undergraduate research advisors:** Who else knows you well, particularly in a research context? The answer is most often an undergraduate research advisor or a post-bach research advisor. This person (or persons, it's ok to ask more than one) probably already wrote a letter than successfully got you into a graduate program and will likely edit that. And, given that it got you into grad school, it will not likely raise any red flags.

- **A member of your guidance committee:** Particularly if you are applying while a second year graduate student or later, you likely have some faculty at your graduate institution that can speak about you. If you've formed a guidance committee, then the people on this committee would be good to consider. They know you, have some research interest that led you to select them for the role, and have some insight into your progress that they can speak to.

- **A graduate program director:** For the NIH applications, the graduate program director speaks very factually about your progress in the program in the Institutional Environment and Commitment to Training section. If your graduate program director does not know you well, then they might not have more to say beyond this. However, if you have had additional interactions with this person, they may also be in a position to provide a letter that speaks to, perhaps, volunteer work you've done for the program, resilience if you faced a challenge getting started in your first year, your thoughtful approach to choosing rotation projects, etc.

- **A professor from a class:** If you consider a professor from a class, be sure it is one who knew you because of the small class size or some particular effort on your part. Do not choose someone who taught a class you liked but perhaps did not know much about you to know what their letter might reflect. But, chances are you took a class in your first year/semester as a grad student and that person could describe your contributions to the class and comment on your interest and scholarly performance. Alternatively, you may have asked a professor from an undergrad course to write a letter for your graduate school applications and may also provide a relevant letter. I advise this to be a '4th or 5th' letter – i.e., lower on the list of priorities.

On the whole, be sure to have letter writers that speak to:
- Your incredible potential as a scientist!
- Your motivation, drive, persistence, determination
- Your experiences preparing you for a career in science
- Your intellect
- Your potential for other aspects of the career that you aspire to.
- Your basic skills like writing and speaking. You may aspire to be better at these, but letters supporting accomplishments in these areas are helpful.
- Your specific lab skills (e.g., 'has quickly learned the techniques in my lab')

Most important is that you give your letter writers a significant heads up as well as reminders as the time gets closer. A suggested timeline is as follows:
- 10 weeks from the deadline: Email letter writers asking if they are willing
- 8-10 weeks (upon response): Send relevant information.
- 4 weeks (if not yet submitted): Send reminder 1
- 2 weeks (if not yet submitted): Send reminder 2

Your request should provide four things:

1) A polite request that they provide a letter of recommendation for you
2) A timeline of when it is due and when you would send reminders
3) Suggestions as to what their letter could speak to. Importantly, reviewers will rate you on (1) academic record and research experience and (2) potential for a successful career as an independent researcher in the future. Let your letter writers know that this so they can <u>specifically</u> speak to these points in the letter.
4) An attachment with NIH recommendations for recommenders (with your specific information included at the top)

An example of an email is as follows:

Dear Prof. UndergradAdvisor,

I hope this finds you well. I am still really enjoying the Microbiology graduate program here at the University of Springfield. It's been an exciting first year for me.

I am currently preparing an application for a NIH pre-doctoral fellowship (NIH F31). I am writing to inquire whether you would be willing to provide a letter of recommendation for me for this application? The letter is due to NIH on Dec 5. As I confirm my application materials, it would be helpful if the letter was submitted 2 weeks in advance of that date, Nov 21.

I thought your letter could reflect the following:
- My research potential – including techniques that I learned, working independently, and my curiosity and exploration through my honors thesis process.
- My experience with presentations – both the poster I presented at The Science Conference and based on talks I was able to give in lab meetings during my time in your lab
- Your letter may also reflect my experience TA-ing for you, demonstrating my ability to both understand Microbiology but also teach it to others.

I am grateful for your consideration of this request. I look forward to hearing back from you.

Kind regards,
Grad Joe

What follows is the instructions for letter of reference writers provided by NIH that you should provide your letter writers, after filling in the top, as an attachment. For a link to the document, go here:

https://grants.nih.gov/grants/how-to-apply-application-guide/submission-process/reference-letters.htm

Instructions for Fellowship Applicant Referees

Name of Fellow *(First & Last Name as shown in the eRA Commons)*: _____

Fellow's eRA Commons Username: _____
FOA Number: _____

The fellowship applicant is applying for an individual fellowship award. The purpose of this award is to provide support to promising applicants with the potential to become productive, independent investigators in scientific health-related research fields relevant to the missions of participating NIH Institutes and Centers, and AHRQ.

Please put the name of the fellowship applicant at the top of the letter. Also, be sure to include your name and title in the letter.

In two pages or less (PDF format), describe the qualities and potential of the fellowship applicant for the research training for which support is being requested (predoctoral, postdoctoral, or senior fellow). This should include your evaluation with special reference to:

- Research ability and potential to become an independent researcher
- Adequacy of scientific and technical background
- Written and verbal communication abilities including ability to organize scientific data
- Quality of research endeavors or publications to date, if applicable
- Perseverance in pursuing goals
- Evidence of originality
- Need for further research experience and training
- Familiarity with research literature

Referees may provide any additional, related comments that they believe will help reviewers evaluate the merit of the fellow's application.

Submitting Reference Letters

Letters must be submitted directly to the eRA Commons at: https://public.era.nih.gov/commons/public/reference/submitReferenceLetter.do?mode=new.

Watch a demo on [Submitting Reference Letters through eRA Commons](#).

Reference Letters are due by the application receipt deadline date, but may be submitted any time after the FOA opens. Reference Letters can be submitted before the grant application submission, and will be held and later linked to the appropriate application once they are received at NIH.

You will be requested to enter the following information on-line at the time of submission:

Referee Information:
- Referee First Name (Required)
- Referee Last Name Required)
- Referee MI Name (Not Required)
- Referee e-mail (Required)
- Referee Institution/Affiliation (Required)
- Referee Department (Required)

Fellowship Application Information:
- PD/PI (Fellowship applicant) Commons User ID (Required)
- PD/PI's Last Name, as it appears on the PI's Commons account (Required) (will be validated to ensure they match)
- Funding Opportunity Announcement (FOA) Number (Required and must match the number of the FOA under which the application is being submitted)
- Reference Letter Confirmation Number (Required only if resubmitting a letter; not required otherwise)
- Fellowship Letter of Reference – two pages maximum. Must be in PDF format. Letter can be printed, signed, and scanned to create the PDF, but do not add a "digital signature" to the document. Additional tips for creating PDF files can be found at http://grants.nih.gov/grants/how-to-apply-application-guide/format-and-write/format-attachments.htm.

After you have submitted your Letter of Reference, both you and the applicant will receive a confirmation of receipt by e-mail. Your e-mail confirmation will include a Reference Letter Confirmation Number. The Confirmation Number will be required when resubmitting reference letters. Please print the confirmation e-mail for your records.

LETTERS OF SUPPORT

The letter of support gives you an opportunity to "prove" that you have access to the facilities, populations or animal models, or specific expertise outside of what is available directly in your sponsor or co-sponsor's lab.

In many cases this will be provided by another faculty member (whose lab you will draw from). However, in some cases, this letter may come from someone in industry (e.g., vouching that they will be able to provide a unique mouse strain) or a non-faculty university member (e.g., a director of a core facility vouching for rates or personnel support).

The letter of support is relatively short and straight forward so it may be useful to provide sample text to the letter writer. The content will vary depending on what support is being provided. However, a general structure might look like:

<u>Support to provide mentoring in an area:</u>

Dear *GradStudent Name*,

I am pleased to have the opportunity to assist you in your proposed F31 and, [INSERT TITLE]. As you know, I have been studying [AREA OF EXPERTISE]. My work considers [DETAILS ON RELAVENT AREA OF EXPERTISE]. For this reason, I am not only very interested in your research topic, but I can be an asset to you in helping you understand [DATA RELAVENT TO EXPERTISE] and help you interpret your findings from [DATA SOURCE]. [INSERT/EDIT TO EMPHASIZE HOW THIS EXPERIENCE/KNOWLEDGE CAN HELP WITH THE WORK (PERHAPS IN DESIGN AS WELL AS ANALYSIS)]

I am happy to support you through one-on-one meetings to discuss issues and questions you have as you plan your study conduct the work, and analyze this research. I will also direct you to other resources that may be available such as local seminars and colloquia. Additionally, I have committed to serving on your guidance committee and look forward to these group-based mentoring interactions as well.

***

I have the highest enthusiasm for your project and look forward to working with and learning from you.

Sincerely,

Professor Smart

Support to provide materials:

Dear *GradStudent Name*,

I am pleased to have the opportunity to assist you in your proposed F31 and, [INSERT TITLE]. Specifically, I will provide [MATERIAL TO BE PROVIDED]. As you know, my lab studies [AREA OF EXPERTISE]. As such, we have the [CAPACITY OR ACCESS TO CRITICAL RESOURCE TO PROVIDE THE MATERIAL]. For this reason, I am happy to provide [MATERIAL] and help as you implement [USE OF MATERIAL]. [INSERT/EDIT TO EMPHASIZE ANY ADDITIONAL SUPPORT THAT MAY BE NEEDED FROM THIS PERSON IN IMPLEMENTING THE MATERIALS]

[INSERT INFORMATION ON THE TIMING OR RESTRICTIONS ON MATERIALS TO BE SUPPLIED]

I have the highest enthusiasm for your project and look forward to working with you.

Sincerely,

Professor Smart

CHAPTER THIRTEEN
Sponsor/Co-sponsor Requests

First, keep in mind that **a co-sponsor is not required** and many proposals do not have one. Do not feel tied to having a co-sponsor but if your work at all bridges labs or extends at all outside of your sponsor's area of expertise, consider whether there is someone who couple play a formal role, enhancing the feasibility of that aspect of the work. For more on this, see Chapter 6 (Selection of Sponsor and Institution).

The sponsor and co-sponsor need to provide the following materials:
- Sponsor/Co-sponsor Biosketches
- Sponsor/Co-sponsor Statement

Be sure to give these individuals plenty of advanced notice so they can prepare the documents and customize them to your proposal. But also consider providing as much support for them as possible as described below.

More than likely your sponsor and co-sponsor have an NIH-formatted biosketch already. Your request should be that they provide an updated biosketch for your proposal. Assuming they have a biosketch already, this should require that they simply:

(1) Update the personal statement to be specific to your proposal. This should include specific reference to you by name and to your proposed research. Emphasize:
- Fit between the sponsor's research and the candidate's research proposal and interests.

- Previous mentoring experience here - if they have a history of mentoring graduate students and/or others, explain this here.
- If you have a team of co-sponsor, consultants, etc to fill in gaps in your sponsor's support, emphasize this here

(2) Update any awards, publications, or grants. Grants are particularly important to have up-to-date so that they matched what is reported in the Sponsor Statement and demonstrate an ability to support your project financially.

If your sponsor or co-sponsor does not have an NIH-formatted biosketch, you should provide them the link to the template (here: https://grants.nih.gov/grants/forms/biosketch.htm)

Pulling together the Sponsor Statement can be time-consuming for a busy professor who has not done one before. Even for those that have, it may be tempting (explicitly or not) to take shortcuts in preparing this document. For this reason, I recommend preparing this as much as possible for your PI. There are certainly aspects that you have no knowledge of, but providing a template for them to fill in will at the very least speed up their response on completing this document.

> **Format requirements:**
> 6 page limit (total – for sponsor + co-sponsor statements)
> Standard format (.5 in margins, 11-inch Arial font)

Formatting the document

Start by formatting the document for the sponsor. Set the margins and the font size.

Then insert a heading structure as follows:

SPONSOR STATEMENT
A. Research Support Available

Create this table for your PI to fill in:

Table 1. Current and Pending Support for Sponsor, [SPONSOR FULL NAME]

Grant Title (Role on Project)	Funding Source and Grant ID	Project Period	Current Year Direct Costs
Active			
Pending			

B. Sponsor's Previous Fellows/Trainees

Create this table for your PI to fill in (the first line provides an example as to how it might be filled in):

Table 2. Past Fellows/Trainees for Sponsor, [SPONSOR FULL NAME]

Trainee	Starting credentials	Training program/degree	Project title	Funding support	Current position
Sally Ride, PhD	B.S., Biology, Springfield University	Biology, PhD	The role of sleep on memory consolidation in platypus	NIH R01 HL5551212	Assistant Professor, Treeham State Univ

C. Training Plan, Environment, and Research Facilities

<u>Individual Development Plan</u>

<u>Training Environment</u>

<u>Research Facilities</u>

D. Fellows/Trainees to be Supervised by Sponsor

Create this table for your PI to fill in:

Table 3. Fellows/Trainees to be Supervised by Sponsor, [SPONSOR FULL NAME]

	Post-doctoral trainees	Doctoral trainees	Undergraduate mentees	Other lab staff
Number to be supervised				

E. Applicant's Qualifications and Potential for a Research Career

This may be the most important component. Remembering that this section also serves as the sponsor's letter of recommendation, this is where that typical content would go. In other words, in this section, the sponsor should *RAVE* about the applicant - the applicant's experiences and the applicant's potential (potential to be an awesome independent research scientist).

This is also a section to address gaps - if the student had a week grade in a *relevant area, less experience than other candidates, or some other gap - the sponsor should admit these and provide a positive spin.*

Develop content

Once you have formatted the document, look for sections which you could draft. Perhaps your PI lists their funding on their website and you can build Table 1. Perhaps you are aware of some recent grads and can complete that Table 3. Even if you only know a few, getting this started gives your PI less to do and a clear template for when they have time to work on it.

Section C presents opportunity for you to contribute to as well. For instance, you can help develop the Individual Development Plan as this should align with your Background and Goals plans. Here you should have a very detailed plan for training. The content should reflect that an Individual Development Plan has begun to be developed, some explanation of this content, and also how often you will meet to assess progress. A good resource for developing an IDP is here: http://myidp.sciencecareers.org/

For Training Environment, you will describe opportunities available to trainees such as yourself. This could be broken into categories such as:
- <u>Lab environment</u> (in this section, discuss how you use lab meetings, journal clubs, and one-on-one meetings to support the trainee)
- <u>Departmental/program environment</u> (here describe courses, mentoring, or other opportunities provided at the department or graduate program level)
- <u>University environment</u> (here describe opportunities for professional development through the Grad School or other opportunities on campus for you to develop grant writing and/or teaching skills)

Although you cannot do much for Section E, it is worth providing the mentor with some bullet points that they might address in this section.

CHAPTER FOURTEEN

Institutional and Environmental Commitment

The purpose of this section is to describe your formal graduate training program, your progress in the program, and other opportunities offered that would support your training. The **good news** is that this section will be completed by your department chair or graduate program director. However (the bad news?), I strongly recommend providing them the 'bones' to both expedite receipt of the document but also guide that person to provide as much information as possible (and accurately).

> **Format requirements:**
> 2 page limit
> Standard format (.5 inch margins, 11-point Arial font)

Although this will be officially from, and signed off by, your graduate program director or department chair, you can draft it.

Graduate training program: This section describes the bones of the program as you might see in a brochure.
- Structure of the program: Start by providing a description of the graduate training program. Consider covering:
  - Structure: If it is a departmental graduate program, is it one of many graduate programs in a department? Or perhaps it is an interdisciplinary (interdepartmental program) in which case you could describe the departments involved.

- o Administration: How is your program run? Often the answer is that there is a graduate program director with a number of committees (e.g., an executive committee, graduate studies committee) who support the program.
- Courses offered: This should provide both the courses offered in the program as well as the coursework requirement (which may be courses outside of your program). A table illustrating the timing of these courses is useful.
- Milestones in training: This is where you describe what the milestones in your program are. These are typically requirements like a quals exam, a comps exam, a dissertation proposal, and a dissertation defense. If possible, indicate the approximate timeline for these activities (e.g., 'end of 2nd year')
- Teaching requirement (optional): particularly if your program has a required teaching component (for instance, TA-ing for 1 semester), explain this requirement here.
- Average time-to-degree: This may be a number you cannot provide but be sure to indicate where the program director might fill in for you the average time to degree for students in your program.

Intellectual environment: This section describes opportunities offered by the program or affiliated with the program. Consider topics such as:
- Seminars - Describe the frequency with which they meet. Emphasize speakers and provide relevant examples.
- Journal clubs - Describe the group offering the journal club, the frequency it meets, and the format.
- Campus-hosted conferences - Describe the topic, the audience, and frequency of such conferences.
- Program retreat - Does your program have an academic retreat? What is the format? How frequently does it occur?
- Travel grants or funds - Describe any support your program offers for conference or other academic travel.
- Career training opportunities - Describe any workshops or efforts your program has towards career development.
- Ethics training - If your program offers any courses or seminars in ethical conduct of scientific research, describe that.
- Facilities and other resources - this is a chance to emphasize the facilities you have available that create a good academic environment.

A note on this section: This will overlap with the content you have in your Facilities and Resources section. That is fine. In this section it is given an overview of the academic environment. In Facilities and Resources you can provide much more detail.

Applicant's progress towards degree: This paragraph should be a statement of your progress in meeting your program goals. For instance:

"X has successfully completed the first year coursework and is enrolled for X for fall 2019. He will take the exam portion of the Comprehensive Exam in January 2019. He has completed a 10-hr/week teaching assistantship, completing his teaching requirement.

Student progress is reviewed by the Graduate Operating Committee each spring. X is both in good standing and on track for success."

Summary and sign-off: This section can end with a summary paragraph and then be signed by the graduate program director. For instance:

"In combination with the training support of X's mentor, X, and faculty in our program, X will have ample support. He will have the intellectual environment to stimulate his understanding of the field, he will have the research tools to accomplish his goals through the Institute of Sciences and X's lab, and he will educational opportunities for preparing for his career ahead through our graduate school. We look forward to supporting him and watching his successes.

Provided by XX
Graduate Program Director
Science Program
University of Brooklyn"

CHAPTER FIFTEEN
Respective Contributions

The objective of this section to tell the reviewers how you and your mentor work together – both in developing the proposal and as you move forward if it is awarded. If you will work closely with another faculty member other than your PI, it is recommended that you include them in your training plan as well.

The ideal situation to reviewers is that you are quite the genius. Their ideal is that you came to this lab with a great idea to combine something from your undergrad experience with what your grad school PI does that will make a significant research advance. They would love it if you genuinely came up with the idea and the PI added something and shined up the text and now you have this work of genius.

But, that is rarely the case. In reality, you may have an idea from you PI that you have developed. It's important to be honest. So the best thing you can do in this case, is be sure you are taking ownership of the project. Are you writing independently? Doing sufficient reading to develop the idea? Are their components that you have added or could add based on your reading or experience? Consider ways that you can still (even though the clock is ticking!) tell the reviewers that this is honestly a mentored project of yours.

> **Format requirements:**
> 1 page limit
> Standard format (.5 inch margins, 11-point Arial font)

Here I recommend breaking this down as follows: a paragraph on your role in developing the idea, a paragraph on your PI's role in developing the proposed work, then a paragraph looking forward.

- <u>Role of grad student applicant</u>: In this paragraph, start with a 1-sentence statement of the project followed by a statement of how you came to be interested and involved in this work. <u>It is important to emphasize your unique contribution here.</u> From there, you can discuss what you have done to this point to support your proposal.

- <u>Role of the PI and other contributors</u>: In this paragraph, give credit to what your PI has done to support you as you developed the idea and the proposal. If they had the base idea (perhaps it was part of their grant proposal), how have they mentored you as you've taken it on? How has your work evolved to be sufficiently related and sufficiently unique? If you worked with multiple faculty, be sure to include their role here as well.

- <u>Future contributions for the proposed work</u>: If the project is funded (and theoretically, even if not), how will you and your PI continue to work together, with you taking the lead on the project but with you PI and any other contributors aiding in the process? Here it is wise to include your Guidance Committee's role as well.

- <u>Table of Roles, Contributions, and Approach to Contributions</u>: Particularly if you have a Guidance Committee that can assist with your topic, it is good to put this into a table format to spell out the respective contributions. Remember, you can use a smaller font in a table, so this allows you to spell out the roles in a bit more detail.

* * *

RESPECTIVE CONTRIBUTIONS

Role of Mia Goodscientist: The proposed research is based on an idea that I generated following a lab discussion of the work by Smith and colleagues (Smith, Jones, & Chambers, 2017) that this project is based on (see Research Strategy). Specifically, based on the first-year project I was doing in Dr. Science's lab, I saw our technique of XX as providing a unique opportunity to address what they recognized as a gap. With encouragement from Dr. Science, I continued to explore this literature and developed the general approach as presented here. In short, we will *brief description of goals*. This project presents an opportunity to *brief statement of relevance*.

Role of Dr. Mad Science and Dr. Good Psych: Dr. Science was critical in developing this proposal, providing the training in the *important technique* and pointing me to relevant literature. More specifically, Dr. Science was interested in the Smith and colleagues paper because he saw the work in our lab going in a similar direction. As such, Dr. Science has provided feedback on many iterations of the study design over the course of many meetings.

As we recognized the need for *critical ancillary area* in the design, Dr. Science encouraged me to meet with Dr. Psych to properly integrate this in the protocol. Dr. Psych has given me feedback on the design and I have recently begun training in *crucial technique*. under his graduate student.

Both Drs. Science and Psych have read drafts of this proposal and provided supportive feedback. These drafts have helped me develop my expertise in this area as well as given me critical grant writing skills.

Future Contributions for the Proposed Work: I will be the Principal Investigator of the proposed research. My role will be to collect data, analyze data, provide summaries and interpretations, and write the work for publication. I will also take advantage of opportunities to present that data at regional and national conferences. Dr. Science will continue to support me as my primary mentor. We will continue to have weekly meetings in addition to any meetings needed as issues arise. I also learn from him via weekly lab journal clubs (where we discuss recent findings from our field), lab meetings (where the lab staff discuss their respective projects), and other informal interactions.

Dr. Psych will serve as a secondary mentor, particularly with respect to the critical technique aspect of the work. I will schedule meetings as needed when I am preparing this portion of the work. Additionally, I will attend lab meetings for his lab when possible.

Additionally, I will utilize members of my guidance committee – Dr. X, Dr. Y, and Dr. Z. – to provide mentoring in other areas of my training (see below).

Team Member	Role	Contribution	Approach to contribution
Mia Goodscientist	PI	I will collect and analyze data, provide summaries and interpretations of the data, and draft the subsequent manuscripts.	I have developed the described team-based approach to my research and career development plan.
Dr. Science	Sponsor	Dr. Science will oversee the technique#1 data collection aspect of this project and assist in data interpretation.	We will meet weekly. Additionally, I will attend lab journal clubs and lab meetings to develop my understanding of this field.
Dr. Psych	Co-sponsor	Dr. Psych will oversee the technique#2 aspect of this project. He will aid in my training on this technique and review data and assist in interpretation.	When this data is collected and analyzed, I will meet weekly with Dr. Psych. I will also attend his lab meetings to gain a broader understanding of this field.
Dr. X	Consultant (Guidance committee member)	Dr. X is an expert in the area of XXX.	We will schedule meetings as needed to discuss critical readings and other interpretations of X.
Dr. Y	Collaborator (and Guidance committee member)	Dr. Y will mentor me in the use of NEW TECHNIQUE. He also agree to provide access to IMPORTANT EQUIPMENT.	We will bi-weekly initially and more frequently when X data is being collected.
Dr. Z	Guidance committee member	Dr. Z is an expert is scientific writing and will aid in my development as a writer.	I will take Dr. Z's course on Science Writing for Grad Students. Additionally, she will mentor me by providing feedback on my manuscripts.

CHAPTER SIXTEEN
Proposal Summary/Abstract

Importantly, the Proposal Summary is what is entered into the NIH RePORTER if the application is funded. This tells you a few things:

(1) **Do not** say anything that you do not want very widely available. Of course this is a reminder to not make grandiose claims that you cannot support. But it is also reason to not make statements regarding anything highly confidential in your work that may be quickly scooped or published before you will.
(2) **Do** describe your work at the level at which a general audience can understand it.
(3) **Do** give the broad view of your project
(4) **Do not** give specific details on methods
(5) **Do not** use personal pronouns of any sort. Write as if writing about someone else's work.

> **Format requirements:**
> 30 lines of text
> Standard format (.5 inch margins, 11-point Arial font)

Thirty lines is not much space to fill, about a half page. This section does not get the attention from reviewers that the Specific Aims and Research Strategy do, so I recommend not 'overthinking' this section. At the same time, reviewers will notice inconsistencies, so it is worth being careful.

***

A general formula is as follows. Note that much of this text might be copied and pasted from your Specific Aims, but be sure to smooth it out after you do.

1-2 sentences: Interesting-grabbing statement and public health relevance
   1-2 sentences: current knowledge
   1-2 sentences: gap in current knowledge
   1 sentence: long-term goal
   1 sentence: specific objective of this proposal
   1 sentence: central hypothesis behind this proposal
   1-4 sentences: statement of specific aims depending on how many you have
   1-3 sentences: some reference to the approach (should be enough to identify model system, techniques to be used)
   2 sentences: statements regarding significance

Examples of Proposal Summaries

You can easily locate examples similar to your research field by
going to: projectreporter.nih.gov
scroll down to Project Details
In the "Project Number/Application ID" just enter "F31"
Press enter
Once you see the list, you might sort by the "Funding IC" column and focus on those Institutes where your research is to fall under (e.g., National Institutes on Aging (NIA), National Institute of Mental Health (NIMH), etc). You can find the acronyms here: https://grants.nih.gov/grants/acronym_list.htm

Project Summary/Abstract Age-related cognitive decline gradually devolves into dementia (e.g. Alzheimer disease, AD). Emotional and healthcare burden, and the fact that AD-neuropathology precedes cognitive changes by many years, make the identification of biomarkers of early disease progression critically important. Aspects of AD neurochemistry other than β-amyloid need to be considered while searching for a reliable biomarker. Glutamate (Glu), the primary excitatory neurotransmitter involved in cognitive processes, is reduced in several key brain regions (specifically the hippocampus, HC) in AD and normal aging. Episodic memory decline is the first cognitive symptom in AD, as well as normal aging. Episodic and associative memory relies on hippocampal Glu. Understanding age-related variations in this system may help track early decline in cognition. Proton functional magnetic resonance spectroscopy (1H fMRS) is the only non-invasive neuroimaging technique that can detect in vivo levels of Glu. We have recently demonstrated that 1H fMRS can detect the temporal dynamics of hippocampal Glu in healthy young adults, which in turn can predict learning proficiency. The long-term goal is to better elucidate the contribution of the glutamatergic system underlying age-related cognitive deficits. The overall objective is to demonstrate that 1H fMRS assessment of task-dependent changes in brain Glu can be harnessed for early prediction of the impending decline in a brain system central to AD. The overarching goal of this proposal is to investigate the relationship between Glu modulation and memory efficiency. The central hypothesis is that baseline levels of Glu, and Glu modulation, will be lower in the elderly, and will be positively correlated with their performance. Guided by strong preliminary data, the hypothesis will be tested by pursuing the following specific aims: 1) Identify the effect of age and basal Glu on associative learning/memory, 2) Determine the effect of age on task-related Glu-modulation in the HC; 3) Investigate whether age-differences in hippocampal Glu modulation during encoding are related to those in learning efficiency. Under these aims 1H fMRS, determined a feasible technique in the applicant's lab, will be used and extended to healthy old adults. Analyses will be conducted to identify a relationship between age-differences in Glu modulation and memory performance. Extant studies examining age-effects on Glu have done so using a non-task-active, static approach. The innovative approach here suggests the utility of variation in task-mediated Glu modulation as an early harbinger of age-related cognitive decline. The proposed research is significant, as it is expected to contribute to, and advance, the search for reliable functional neural biomarkers of neuro-degenerative disorders, and help track cognitive changes. Investigation of age-effects on the dynamics of Glu, an important neurotransmitter, will help gain new insights on age-differences in neurotransmission capacity and provide an effective framework to test better-targeted therapies to mitigate impending cognitive decline.

From my own R01:

PROJECT SUMMARY

Sleep protects and enhances memory in young adults: performance changes on a range of tasks are greater following an interval with sleep relative to changes over an interval spent awake. In young adults, a mid-day nap is sufficient for gaining these performance benefits. Unlike adults, mid-day naps are routine for young children. As such, the nap opportunity in preschools may serve as a target for intervention in children with learning impairments and reduced overnight sleep opportunities. **However, whether naps confer a particular benefit to learning and performance of young children is unknown.** The <u>specific objective</u> of the proposed research is to examine whether naps contribute to immediate and delayed benefits on multiple forms of learning in young children (3-5 yrs). By probing recall prior to and following mid-day nap or wake intervals, the <u>overarching hypothesis</u> is that recent memories are actively processed (as opposed to passively protected) by a nap, conferring immediate and delayed (24-hrs) benefits on declarative (Aim 1), procedural (Aim 2), and emotional (Aim 3) memories. In two conditions, children will either be nap-promoted or wake-promoted mid-day. Subsequently, performance will be reassessed that day as well as the following day. The specific hypotheses examined are: a) mid-day naps benefit learning; b) naps yield stronger memories at 24-hrs; c) performance benefits are due to an active role of sleep as indicated by a relationship between sleep physiology and behavior. <u>This work is innovative</u> in that it presents a novel application of an accepted theoretical construct. Moreover, these results are expected to shift the current practices regarding naps in preschools to a practice of nap-promotion and better regard for the length of the nap opportunity. The **translational significance** may be seen in new policies regarding in-class nap opportunities and pediatric nap guidelines for preschool children. The **theoretical significance** is that these outcomes will drive an entirely new research dimension for educational sciences (sleep as a novel target to enhance learning) and spur further developmental studies on the influence and underpinnings of sleep-dependent cognitive and neural processes.

CHAPTER SEVENTEEN

Human Subjects

If you are using human research participants, the human subjects section is required. Although this is typically self-evident, when in doubt, discuss with your sponsor whether these sections will be required for you.

The Human Subjects section will require information similar to what you provide in an human subjects /institutional review board (IRB) protocol. The goal of this section is to assure reviewers that you have considered the ethical issues in the use of human participants in the research you plan.

It is important to put great care into this section. Reviewers tend to scrutinize this section of graduate student proposals in particular – if this section is not sufficiently thorough, this is interpreted as a lack of mentoring/communication with your sponsor and/or that you did not think through what your project will require in sufficient depth.

> **Format requirements:**
> **no** page limit (1.5-3 pages would be a reasonable length)
> Standard format (.5 inch margins, 11-point Arial font)

1. Just as in the other sections, start with formatting your page (see above) and putting your headers in place:

PROTECTION OF HUMAN SUBJECTS

* * *

1. **Risks to Human Subjects**
a. Human subjects involvement, characteristics, and design
b. Sources of materials
c. Potential risks
>Physical
>Psychological
>Financial
>Social
>Legal

2. **Adequacy of Protection Against Risks**
a. Recruitment and informed consent
b. Protection against risks

3. **Potential Benefits of the Proposed Research to the Subjects and Others**

4. **Importance of Knowledge to be Gained**

2. With that skeleton on paper, now fill it in. You can draw off from materials and experiences you have had but if your experiences are limited, read the descriptions below carefully – what is expected from your internal human subjects review committee may have a different focus than grant reviewers.

1. **Risks to Human Subjects**
a. Human subjects involvement, characteristics, and design: This section provides the description of the human subjects in your study. Suggested breakdown:
-
    - Paragraph 1: Number of participants and a description of participants (such as age or groups)
-
    - Paragraph 2 (or more): Inclusion and exclusion criteria

- Paragraph 3: How participants will be recruited (advertising, database, etc)

b. Sources of materials: Consider all of the measures you have in the study. These each result in data. Describe these one by one.

c. Potential risks: Describe risks associated in the following categories. If none, state that that is the case.

- Physical: Will you expose your participant to pain or even discomfort, even if temporary or mild?

- Psychological: Does your research have the potential to cause participants psychological harm, even if unlikely more mild? Might participants feel frustrated or emotional or embarrassed?

- Economic: Is there any possibility of an economic cost to participants?

- Social: Does your research have the potential to stigmatize a social group – based on political beliefs, sexual orientation, religion, or other social grouping?

- Breach of confidentiality: Regardless of the mechanisms in place, there is some risk of confidential information being accessed by someone other than study personnel. Recognize and explain this risk.

2. Adequacy of Protection Against Risks
a. Recruitment and informed consent

- Paragraph 1: Describe how you will recruit your participants (database, advertisements)

- Paragraph 2: Describe informed consent procedures

b. *Protection against risks:* Review each of the above noted risks, one by one, and note how they are minimized to the fullest extent possible.

3. Potential Benefits of the Proposed Research to the Subjects and Others

State the value gained from the research – both to the participant and to others. This excludes any financial compensation. For example, the findings may have the potential to identify a novel target of intervention for disease or disability (societal benefit) and the participant may feel satisfaction for having contributed to the greater good (individual benefit).

4. Importance of Knowledge to be Gained

The outcomes must have significant benefit to justify the risks you describe. So, although you will point to this significance in the Significance section of the Research Strategy, it is important to emphasize this here as well.

CHAPTER EIGHTEEN
Vertebrate Animals

If you are using vertebrate animals, the vertebrate animal section is required. Although this is typically self-evident, when in doubt, discuss with your sponsor whether these sections will be required for you.

The Vertebrate Animals section is similar to animal research/ Institutional Animal Care and Use Committee (IACUC) protocol that you have likely done before. If not, be sure to ask your sponsor to see the protocols used in the lab.

This section is particularly important for a graduate fellowship review – although it will not make a mediocre proposal great, it can make a great proposal mediocre if you seem to not recognize the importance of this section or minimize the potential for harm.

> **Format requirements:**
> no page limit (1.5-3 pages would be a reasonable length)
> Standard format (.5 inch margins, 11-point Arial font)

1. Just as in the other sections, start with formatting your page (see above) and putting your headers in place:

VERTEBRATE ANIMALS
   1. Description of Procedures on Vertebrate Animals
   a. Animal characteristics

b. *Description of animal procedures*

2. **Justification for Use of Vertebrate Animals**
   a. *Choice of species*
   b. *Consideration of alternatives*

3. **Minimization of Pain and Distress to Vertebrate Animals**
   a. *Circumstances leading to discomfort, distress, pain, or injury*
   b. *Procedures to alleviate discomfort, distress, pain, or injury*

4. **Method of Euthanasia (optional)**

2. Now expand each of these sections. Read the descriptions carefully to determine the scope of your sections.

VERTEBRATE ANIMALS
   1. **Description of Procedures on Vertebrate Animals**
      a. *Animal characteristics:*

   b. *Description of animal procedures*

   2. **Justification for Use of Vertebrate Animals**
      a. *Choice of species*
      b. *Consideration of alternatives*

   3. **Minimization of Pain and Distress to Vertebrate Animals**
      a. *Circumstances leading to discomfort, distress, pain, or injury*
      b. *Procedures to alleviate discomfort, distress, pain, or injury*

   4. **Method of Euthanasia (optional)**

CHAPTER NINETEEN
Resource Sharing Plan

The data and any resources generated by your research provides an additional asset to the research. For instance, you might collect a substantial amount of genetic data in search of X, but the data could be made available for someone else to come along and search for Y. How the data will be preserved and shared is of interest to NIH.

This section has no page limit but is typically 1/3-3/4 page.

The Resource Sharing Plan section is standard in NIH grants, R01s and F31s alike. Although the text is somewhat standard across proposals, it should be customized to the type of data you will be generating.

1. Start by formatting the document. Although there may not be a need for section headings, one possible configuration is:

RESOURCE SHARIN PLAN
   1. Dissemination to the scientific community
   2. Dissemination to the lay audience
   3. Data sharing plan

2. Now fill in these sections.

* * *

RESOURCE SHARIN PLAN

Resources generated with funds from this grant will be freely distributed, as available, to academic investigators (for non-commercial research) and the general public. The [UNIVERSITY NAME] will adhere to the NIH Data Sharing Policy (https://grants.nih.gov/grants/guide/notice-files/NOT-OD-03-032.html).

1. Dissemination to the scientific community

Here discuss ways that you will disseminate your research findings to the scientific audience. Sub-headings might be:
- scientific conferences (particularly in the target disease or research area)
- scientific publications
- scientific interest groups (specify newsletter, website, etc)

2. Dissemination to the lay audience

Here discuss ways that the findings, and your research generally, will be disseminated to the lay audience, particularly those relevant to your research area. Some examples might be:
- Annual [DISEASE] Awareness Week
- Support groups
- Newsletter or social media feed of patient-based organization

3. Data sharing plan

Discuss how raw data or other resources will be made available. Although a generic statement can be made (e.g., "Research resources generated with funds from this grant will be freely distributed, as available, to qualified academic investigators for non-commercial research", a more specific statement is preferred. Consider where data (particularly large file sizes) may be stored - both on-campus repositories to public access archives such as Dataverse or re3data.org.

CHAPTER TWENTY
Bibliography/References Cited

NIH does not have a specific reference citation format. There are also no page limits on the Bibliography section. So, you might think that you cannot go wrong. WRONG. In fact, there are many ways to go wrong:

- failing to include a reference to critical research (maybe you cited it and just failed to list the reference, or maybe you didn't cite it at all).
- having mismatched formatting of references
- other errors, particularly large, glaring errors, in the reference format

A few recommendations:

1. Format the section just as all your other pages:
   Insert the title on the page:
   REFERENCES

Set all other formatting (border, font size, arial)

2. Choose a reference format that makes sense. You might choose the format from your major research journals in your field (for instance, if you are a psychologist, use APA formatting for the references).

3. Use a reference manager that will allow you to try different formatting in no time. However, do not trust your reference manager too much - sometimes references are added without volume or page

numbers, with titles in ALL-CAPS, or with other errors.

CHAPTER TWENTY-ONE
Editing your Proposal

As you near the end of your proposal writing, it is important to view the document through a reviewer's eyes. This person will be writing comments based on specific criteria that do not match directly to individual sections, but are found across sections. So, after you have drafted the proposal, review the materials with these criteria in mind:

Fellowship Applicant
What the reviewers will look for:
- Do the materials convey that you have a strong academic record and research experience?
- Do the materials convey you have strong potential for a future career as an independent researcher?
- Have you conveyed a commitment to future research in your materials?

Where the reviewers will look:
- Applicant biosketch: Is your research experience sufficiently explained? Does it 'sell' that you have experience and potential? Does the personal statement emphasize your interest and commitment to future research?
- Background and goals: Here, too, have you emphasized your research experience and potential? Do you convey a clear target for a future as an independent researcher?
- Letters of support: Although you have no say in the content of your letters, this is a key place for reviewers when rating the Applicant. This is why it is important for letter writers to specifically speak to these points.

Sponsors, Collaborators, and Consultants

What the reviewers will look for:
- Does the sponsor have the experience (both research and mentoring experience) to support you?
- Are you and your sponsor a good fit - particularly with respect to your research interests and their research portfolio?
- Does the sponsor demonstrate commitment, including of research funds, space, and resources?
- Are the roles of co-sponsors, collaborators, and consultants specified?

Where the reviewers will look:
- <u>Selection of Sponsor and Institution:</u> Do you convey a good fit between your research interest and what your sponsor does?
- <u>Sponsor Biosketch:</u> Although your sponsor will provide their biosketch and you don't have any say in it, with these materials in mind, you might note a gap that can be emphasized in other materials in this section.
- <u>Sponsor Statement</u>: Does the Sponsor's Statement emphasize their experience as a mentor, sufficient funding to support the costs of your project, and commitment of time, space, and other resources?
- <u>Respective Contributions</u>: Does this section emphasize the sponsor's fit for the project and role of other other team members? Collectively, will the reviewers say "This is a great mix of support for this candidate's career and research goals"?

Research Training Plan

What the reviewers will look for:
- Is the research solid, well-justified, and feasible?
- Does the research align with the candidate's goals?
- Is the research overlapping with <u>but distinct from</u> the sponsor's research?
- It the research feasible - in the time your propose and with the resources you have available?

Where the reviewers will look:

- Research Strategy: Of course the primary area the reviewers will look to in order to justify their score in this area is the Research Strategy. Look back at this section and think about how well you 'sell' the research as significant, innovative, and clear. Will the reviewers find it feasible (see Feasibility section of the Research Strategy)?
- Background and Goals: Does the timeline in the Research Strategy align with the timeline in the Background and Goals section? Are they consistent and clearly feasible? Will the research help you attain your goals?

Training Potential

What the reviewers will look for:
- Is the training plan appropriate and thorough for the career goals?
- Does the plan take advantage of existing strengths?
- Are the gaps addressed through the research and training plans sufficient enough to warrant support? In other words, is it needed? Is it valuable?
- If successful, will the training plan result in the candidate being sufficiently prepared to be an independent researcher?

Where the reviewers will look:
- Background and Goals: In this section, do you state a specific career objective, one of an independent research career? Will the steps you propose address gaps in your training and prepare you for an independent research career? In this case, be sure to think like an outsider - with a clear mind, ask, what gaps might they see in my training in order for me to be independent? You might look at what aspects of research your sponsor seeks consultants for. For instance, do you have sufficient training in statistics, computer programming, and writing? Do you tie your strengths (Background) in with your training plan (Goals)?
- Biosketch (candidate): Does your personal statement emphasize your goal and how this proposal will help you attain it? Does it emphasize your existing strengths (perhaps you have had non-course based training in statistics or

programming in a research lab that you could describe).

Institutional and Environmental Commitment to Training
What the reviewers will look for:
- Are research facilities and equipment adequate (or, better yet, superior) for what you propose?
- Are there ample training opportunities, fitting with your gaps and goals?

www.ingramcontent.com/pod-product-compliance
Lightning Source LLC
Chambersburg PA
CBHW070240220526
45465CB00004B/1466